MIXED-MODE SIMULATION

THE KLUWER INTERNATIONAL SERIES
IN ENGINEERING AND COMPUTER SCIENCE
VLSI, COMPUTER ARCHITECTURE AND
DIGITAL SIGNAL PROCESSING

Consulting Editor
Jonathan Allen

MIXED-MODE SIMULATION

by

Resve A. Saleh
University of Illinois

and

A. Richard Newton
University of California

KLUWER ACADEMIC PUBLISHERS
Boston/Dordrecht/London

Distributors for North America:
Kluwer Academic Publishers
101 Philip Drive
Assinippi Park
Norwell, Massachusetts 02061 USA

Distributors for all other countries:
Kluwer Academic Publishers Group
Distribution Centre
Post Office Box 322
3300 AH Dordrecht, THE NETHERLANDS

Library of Congress Cataloging-in-Publication Data

Saleh, Resve A., 1957–
    Mixed-mode simulation / by Resve A. Saleh and A. Richard Newton.
      p. cm. — (The Kluwer international series in engineering and
computer science. VLSI, computer architecture and digital signal
processing)
    Includes bibliographical references.
    ISBN-13:978-1-4612-8030-9   e-ISBN-13:978-1-4613-0695-5
    DOI: 10.1007/978-1-4613-0695-5

    1. Integrated circuits—Very large scale integration—Design and
construction—Data processing. 2. Integrated circuits—Very large
scale integration—Computer simulation. 3. Computer-aided design.
I. Newton, A. Richard (Arthur Richard), 1951–   II. Title.
III. Series.
TK7874.S23   1990
621.39 '5—dc20   90–34604
                                       CIP

TABLE OF CONTENTS

PREFACE

Our purpose in writing this book was two-fold. First, we wanted to compile a chronology of the research in the field of mixed-mode simulation over the last ten to fifteen years. A substantial amount of work was done during this period of time but most of it was published in archival form in Masters theses and Ph.D. dissertations. Since the interest in mixed-mode simulation is growing, and a thorough review of the state-of-the-art in the area was not readily available, we thought it appropriate to publish the information in the form of a book. Secondly, we wanted to provide enough information to the reader so that a prototype mixed-mode simulator could be developed using the algorithms in this book. The SPLICE family of programs is based on the algorithms and techniques described in this book and so it can also serve as documentation for these programs.

ACKNOWLEDGEMENTS

The authors would like to dedicate this book to Prof. D. O. Pederson for inspiring this research work and for providing many years of support and encouragement. The authors enjoyed many fruitful discussions and collaborations with Jim Kleckner, Young Kim, Alberto Sangiovanni-Vincentelli, and Jacob White, and we thank them for their contributions. We also thank the countless others who participated in the research work and read early versions of this book. Lillian Beck provided many useful suggestions to improve the manuscript. Yuncheng Ju did the artwork for the illustrations. Funding and computer equipment for this research was provided by the Natural Sciences and Engineering Research Council (NSERC) of Canada, the Hewlett-Packard Company, Toshiba Corporation, Digital Equipment Corporation, and the Semiconductor Research Corporation.

CHAPTER 1

INTRODUCTION TO MIXED-MODE SIMULATION

1.1. THE SIMULATION PROBLEM

Computer simulation is used in a variety of different fields to predict the behavior of physical systems whenever it is inappropriate, or too expensive, to build the actual system to observe its behavior. In electrical engineering, circuit simulation is used routinely in the design of integrated circuits (IC) to verify circuit correctness and to obtain detailed timing information before an expensive and time-consuming fabrication process is performed. In fact, it is one of the most heavily used computer-aided design (CAD) tools in terms of CPU-time in the IC design cycle. The popularity of this form of simulation is primarily due to its reliability and its ability to provide precise electrical waveform information for circuits containing complex devices and all associated parasitics.

Detailed circuit simulation has been used extensively for IC design since the early 1970s. However, the ever-increasing number of devices on a single silicon chip has led to development of a number of higher-level simulation tools to cope with the complexity of the problem. These tools include behavioral simulators, register-transfer-level (RTL) simulators, gate-level logic simulators, and more recently, switch-level simulators. These programs have been used to verify circuit functionality and to obtain first-order timing characteristics. Typically, the higher-level tools provide enough information to design working circuits. However, there is still a significant time lag between a functioning circuit and a circuit which meets the design specifications - particularly in the case of high-performance custom integrated circuits. In fact, circuit simulation is the only tool which provides enough detail to ensure that circuits of this type will meet specifications over a wide range of operating conditions.

At the present time, the most popular circuit simulation tool is the SPICE2 program [NAG75]. There are many thousands of copies of this program in use, as well as a number of versions of "alphabet-SPICE" (e.g., HSPICE, PSPICE, IGSPICE) being marketed commercially. All of these programs offer a wide variety of analyses including dc analysis, time-domain transient analysis, ac analysis, noise analysis and distortion analysis. Of these, the time-domain transient analysis is the most computationally expensive in terms of CPU-time. The SPICE program was originally designed to simulate circuits containing up to 100 transistors. However, at certain companies, this program is often used to simulate circuits containing over 10,000 transistors at great expense! The program is accessed over 50,000 times per month at some of companies with a "job mix" that conforms to the 80-20 rule. That is, 80% of the SPICE runs are on small circuits which consume only 20% of the total CPU-time used each month, while 20% of the jobs are very large and consume 80% of the CPU-time used each month. Therefore, the development of fast but accurate simulation methods for very large-scale integrated (VLSI) circuits continues to be an important area of research.

1.2. LEVELS OF SIMULATION

1.2.1. Electrical Simulation

Electrical or circuit level simulation provides the greatest amount of detail. The electrical transient analysis problem in SPICE involves the solution of a system of nonlinear, first-order, ordinary differential equations. These equations model the dynamic characteristics of the circuit for a set of applied input voltages, given a set of initial conditions. The solutions are voltage waveforms at circuit nodes and current waveforms through circuit elements. Usually the designer is interested in only a subset of the entire set of solutions.

Standard circuit simulators use *direct methods* to solve the circuit equations. Briefly, direct methods employ some form of numerical integration to convert the nonlinear differential equations into a set of nonlinear difference equations. These equations are solved simultaneously using the iterative Newton-Raphson method. This involves a conversion of these equations to linear equations and their subsequent solution using a sparse LU decomposition technique [NAG75]. There are two limitations in this approach that make it inappropriate for large circuits. One fundamental problem is that the sparse linear solution dominates the run time for large circuits [NEW83]. The second limitation is that, at each time point, *all* variables in the system are solved using a common time-step based on the fastest changing component in the system. This can be inefficient for both small and large circuits, but it is more significant for very large problems where most of the components are either changing very slowly or not changing at all.

A variety of techniques have been investigated to improve the performance of circuit simulators. Early work in this area included *timing simulation* [CHA75, NEW78B, DEM80], which is a simplified form of relaxation-based circuit simulation, and *tearing methods*, which have been applied to both linear [SAN77, YAN80, SAK81] and nonlinear [RAB79] equation levels. More recently, the relaxation-based approaches have been the focus of intensive research. In particular, the Waveform Relaxation method [LEL82, WHI83] has been implemented in a number of programs including RELAX [LEL82, WHI83], SWAN [DUM86], TOGGLE [HSI85], RealAx [MAR85], MOSART [CAR84] and iDSIM [OVE89]; and Iterated Timing Analysis [KLE83, SAL84], based on nonlinear relaxation, has been implemented in SPLICE [SAL83, KLE84, ACU89], ELDO [HEN85] and SISYPHUS [GRO87].

1.2.2. Gate-Level Simulation

When the complexity of an integrated circuit design reaches the point at which electrical analysis is no longer cost effective, logic or *gate-level* simulation is used. In logic simulation, transistors are usually grouped into logic *gates* and modeled at the gate level. This form of simplification, sometimes referred to as *macromodeling*, can result in greatly enhanced execution speed by reducing the number of models to be processed and simplifying the arithmetic operations required to process each transistor group. Rather than dealing with voltages and currents at signal nodes, discrete logic *states* are defined, and simple Boolean operations are used to determine the new logic value at each node. Boolean operations are generally the most efficient operations available on a digital computer.

A logic simulator that uses event-driven, selective-trace techniques is typically 100 to 1000 times faster than the *most efficient* forms of electrical analysis. It can also provide first-order timing information, including the detection of hazards, glitches, and race conditions. In addition, it can output information regarding any illegal states or conflict conditions that may arise at any node in the circuit. The number of logic states used in a simulation, their meaning, the logic delay models and the scheduling algorithm all have a profound impact on both the speed and accuracy of the results. It is this wide variety of factors that has resulted in the development of such a large number of logic simulators - almost every one addressing a different set of tradeoffs.

1.2.3. Switch-Level Simulation

Recently, switch-level simulation [BRY80] has become the preferred form of logic simulation for MOS digital circuits. In this
approach, the circuit is entirely simulated at the transistor level, rather
than at the gate level. The transistors are modeled as gate-controlled
switches and operate as follows: if the transistor is "ON," it is viewed as
a closed switch and it may transfer a signal value from one node to
another; if the transistor is "OFF," it is viewed as an open switch and is
incapable of transmitting any signals through it. The network is composed of a set of nodes connected by these switches, and the logic value
at each node is determined using this idealized transistor model. Usually
a strength is associated with each transistor switch when in the closed
position to model the conductance of the device. This strength is used
to determine the effective conductance of signal paths from any node to
the supply and ground nodes. The capacitance at each node can also be
modeled using a node strength that is proportional to the size of the
capacitance. Many of the important features of MOS circuits, such as
charge-sharing and bidirectionality, can be modeled using this switch-
level model, although detailed timing information is not usually provided.

A number of researchers have attempted to incorporate timing
information at the switch level at the cost of additional CPU-time.
Simulators that fall into this category are MOTIS [CHA75], RSIM
[TER83], ELOGIC [KIM84], SPECS [DEG84], MOSTIM [RAO85],
CINNAMON [VID86], SPECS2 [VIS86] and iDSIM [OVE88]. Programs such as RSIM treat MOS transistors as linear resistors and compute signal transition times using the Penfield-Rubenstein technique
[PEN81], which is an RC-delay modeling technique. Although the
method is extremely efficient, the overall accuracy of this approach is

limited due to the simplified nature of the delay model. MOSTIM and iDSIM determine the delay directly using lookup tables for delays that are generated during a pre-characterization phase for recognizable transistor configurations. These tables account for factors such as device sizes, loading and input slew-rate. The ELOGIC and SPECS programs compute the delays by using electrically-based table lookup device models. The waveforms are generated as piecewise linear segments using the computed delays. Both approaches provide for variable precision by allowing the user to specify the number of voltage or current levels to be used in the simulation.

1.2.4. Register-Transfer Level Simulation

Register-Transfer Level (RTL) [BRE75] simulation is concerned with circuits described at a higher level of abstraction. Combinational components, (such as gates, multiplexers, decoders, encoders, adders, and arithmetic units) and sequential components (such as registers and counters) may be used in RTL simulators. RTL simulation has been used extensively for data path design. It is used for both the description and simulation of the designs when evaluating alternative architectures. The set of statements describing the circuit operation involves a sequence of register transfers and arithmetic operations that are similar to data-flow descriptions. In the description, related bits of information are usually grouped into ordered sets of words or buses for convenience and to establish logical relationships. Although RTL simulators are widely used to design computers, they do not usually provide information regarding races, hazards, illegal states or critical timing constraints.

1.2.5. Behavioral Level Simulation

Behavioral level simulators [HIL80, INF84, INS84] allow the designer to define arbitrary functional blocks, both combinational and sequential, that can be used in system-level simulation. Two types of blocks may be defined: *structural* and *behavioral*. Structural blocks describe how a number of functional blocks are interconnected. A behavioral block contains a detailed description of the operations to be performed on the inputs to produce the outputs of the block. The statements describing the operations are usually written in a high-level language, typically a hardware description language (HDL), and then translated to a standard programming language format and compiled into the simulator. When the simulator is executed, the operations of the system are emulated. Examples of applications that are appropriate for simulation at the behavioral level are: verifying of the system timing in a CPU; checking a proposed network protocol for a local-area network; and validating the operations in DMA controller sequence.

1.3. MIXED-MODE SIMULATION

The various levels of simulation, described in the previous section, are listed in Table 1.1 from the highest level of abstraction to the lowest level. The relative runtime cost and accuracy of each simulation level is provided for the hypothetical simulation of a 32-bit microprocessor. The reader should notice that the progression from behavioral level to electrical level provides an increase in the accuracy of the simulation at the cost of more CPU-time. A progression in the opposite direction usually allows larger and larger circuits to be simulated for a given amount of CPU-time, or requires less and less CPU-time to simulate a given circuit. However, each level uses less precision in signal representation. This often translates to less accuracy in the results due to modeling limitations.

There are many situations in which only one level of simulation is not sufficient for the simulation of an entire design. One common

Level	Relative Cost	Capability and Accuracy
Behavioral (B)	1	Algorithmic verification, some timing information
RTL (R)	10	Functional verification, some timing information
Gate (G)	100	Functional verification, first-order timing information
Switch (S)	1000	Functional verification, first-order timing information
Timing (T)	10000	Detailed waveform information with variable accuracy
Electrical (E)	1000000	Most accurate form of simulation

**Table 1.1:   Relative Cost and Accuracy of Simulation
for a Given Example**

situation arises in the design of a mixed analog and digital circuit. Logic simulators do not generally have the capability to model analog circuitry, and it is usually too expensive to simulate the entire design in a circuit simulator. A simple example of this is shown in Fig. 1.1 where a clock generator is represented using resistors, capacitors and logic gates. In this case, it would be convenient if a simulator that included both electrical and logic simulation capabilities were available.

Another situation that requires the use of multiple levels of abstraction is to describe designs developed in a "top-down" or "bottom-up" design environment. In both cases, the entire design at any given point in time may be represented at a number of different levels of abstraction. One designer may have the behavioral specification of his/her portion of the design, while a second is completing the detailed gate-level design,

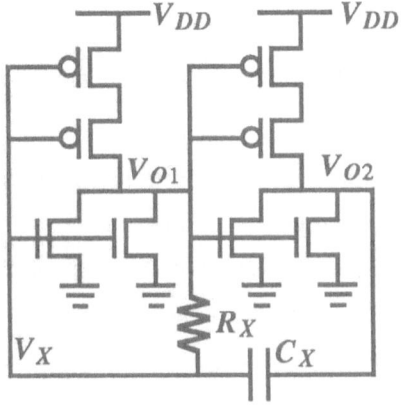

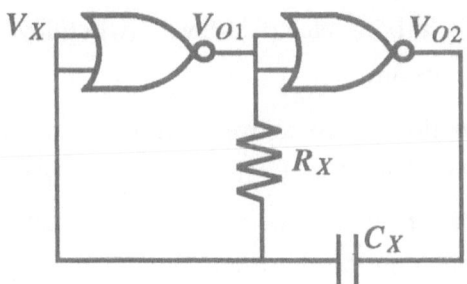

Figure 1.1: A Clock Generator Circuit

and a third is performing transistor-level cell library development. Furthermore, a designer often uses multiple levels of abstraction in a schematic diagram to convey the important aspects of the design as shown in Fig. 1.2. To ensure that the designs represented schematically at many levels is functionally correct at any stage of the design process, a simulator that handles all possible levels of abstraction would be useful.

Mixed-level simulation can also be used for the purpose of accurate circuit modeling. For example, standard gate-level simulators are not capable of simulating the behavior of certain properties of MOS digital circuits such as bidirectionality and charge-sharing. Therefore, the mixing of switch-level simulation and gate-level simulation would provide an effective balance between simulation speed and functional accuracy. On the other hand, the idealized transistor model in switch-level simulation is not appropriate for the simulation of certain pass transistor configurations, and other circuits where the device W/L ratios are important. For these cases, mixing electrical-level, switch-level and gate-level simulations would be useful.

All of the situations cited above require a simulator that allows different portions of the circuit to be described and simulated at different levels of abstraction. That is, where accuracy is not a critical issue, higher levels of simulation can be used, but where proper modeling of the circuit is a problem or detailed timing information is desired, the lower levels of simulation can be used. CAD tools that address this need are referred to as mixed-mode simulators or multi-level simulators, or sometimes mixed-level simulators. This book describes the key issues in mixed-mode simulation and presents enough detail to allow the reader to implement a prototype mixed-mode simulator.

1.3.1. Basic Issues in Mixed-Mode Simulation

In designing a mixed-mode simulator, a number of issues must be addressed. These issues are as follows:

Choice of Simulation Levels: First, and foremost, is the issue of which forms of simulation to include in the simulator. This depends on the intended application of the simulator. If the design is primarily digital in nature, the combination of gate, RTL and behavioral simulations would be appropriate. For MOS designs, it may be better to incorporate gate and switch-level simulations. For designs containing both MOS and bipolar transistors, it may be necessary to mix gate level and electrical level simulations. Ideally, one would prefer to combine all the levels of simulation into one program, but the development time would be significant.

Simulator Architecture: A mixed-mode simulator must be flexible and extensible so that algorithms and device models can be added or removed easily as the technology and the simulator requirements evolve. An appropriate choice of simulator architecture is necessary to achieve this goal. The architecture described in this book is based on the use of the event-driven, selective-trace paradigm at all levels of abstraction. This permits the exclusive simulation of activity, and it is a necessary feature when simulating large digital systems. It is also consistent with the algorithms commonly used in most simulators, except for standard electrical simulators, which must be modified to fit within the event-driven framework.

Event Definition and Event Scheduling: To establish event-driven, selective-trace simulation, the notion of an *event* must be defined at each

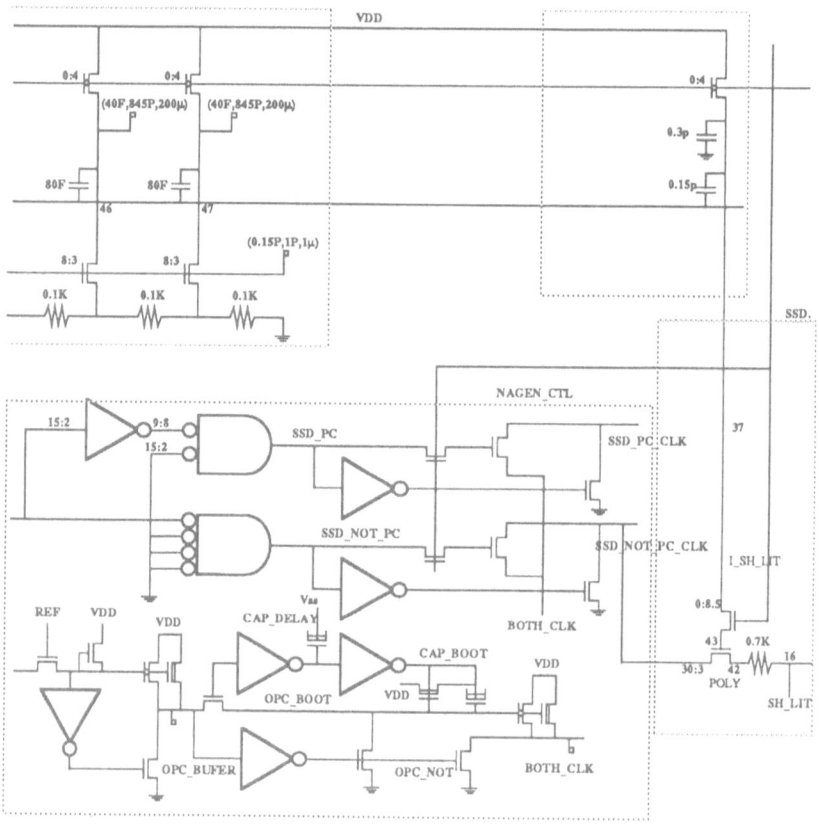

Figure 1.2: A Portion of a Control Circuit

level in the simulator. An event is a change in state of some node in the
circuit that may affect other components in the circuit. The effect of an
event is to cause all fanout components to be processed, and possibly
new events to be scheduled, if changes in their output nodes occur. The
key issue in mixed-mode simulation is to define a scheduling policy for

events that occur between different modes of simulation.

Signal Representation and Signal Mapping: A consistent representation for signals over all simulation levels is critical for accurate mixed-mode simulation. In the higher levels of simulation, the signal value is usually represented using hexadecimal values for collections of bits or logic values such as 0, 1 and X for single bits. At the other extreme, electrical simulation uses 64-bit double precision words to represent real values of voltage. A mixed-mode simulator must be able to manage these different signal types and map them from one representation to another without a significant loss in accuracy.

Representation of Time: Time is usually represented as real numbers in electrical simulation and as integers in logic and higher level simulations. Typically the time steps chosen in electrical simulation are very small (order of nanoseconds to picoseconds), whereas in the logic level and higher levels of simulation, it is usually an integer multiple of some basic unit of time (order of nanoseconds). This disparity between the various representations of time must also be resolved in the mixed-mode environment.

Partitioning: Circuit partitioning is a key factor in obtaining efficiency and accuracy from mixed-mode simulation. The main question is to determine which portions of the circuit must be simulated at the most detailed level and which portions will profit from simulation at higher levels of abstraction without any noticeable loss of accuracy. The prospects of performing this task automatically seem formidable and this is still an "open" research area. To date, most of the simulators available require that circuit designers be responsible for the partitioning process, since they are familiar with the nature of the design.

User Interface: Another important consideration when designing a mixed-mode simulator is the user interface to the simulator. The interface must be graphics-oriented, highly interactive, and provide the features of schematic capture, simulation control and output post-processing. While a variety of schematic packages with these features do exist commercially, there are a number of additional requirements in mixed-mode simulation. First, the front-end must allow a hierarchical representation of the circuit in which each successive level of the hierarchy implies a different level of abstraction. That is, each level in the hierarchy represents a different form of simulation in the associated mixed-mode simulator. This implies that all of the components representing the circuit at two or more different levels must have the same functional behavior to guarantee correct results. Therefore, some convenient way of verifying the consistency of different representations of the same circuitry must be provided by the user interface. The capability of adding new components, specifically, macromodels or high-level RTL and behavioral models, should also be made simple.

1.3.2. A Survey of Existing Simulators

Mixed-mode simulation has been gaining in popularity over the last few years; as a result, a large number of mixed-mode simulators have been developed. The mixed-mode simulation techniques used in the programs can be broadly classified into three groups:

1) **Manual approach (M):** In this approach, a logic simulator is used to simulate the digital portion of the circuit. The results are used as inputs to an electrical simulator to simulate the analog portion. This process is both tedious and unreliable, especially if feedback paths exist

between the analog and digital portions of the design.

2) "**Glued**" **approach (G):**   In this case, two or more existing simulators are combined using either a procedural interface, if the programs are executed in the same address space, or an interprocess communication mechanism, if the programs are running in different address spaces. This is an effective solution for companies that have already invested large amounts of time and money maintaining separate simulators and are not willing to abandon them in favor of the development and support of a completely new simulator. In addition, the input languages for the simulators do not have to be modified and, therefore, have minimal impact on the designer. However, this simple solution also has a number of inherent limitations in terms of efficiency. The processing of bidirectional elements connected across the mixed-mode interface presents a problem, and the time advancement, backup and synchronization of the various simulators that are running concurrently must be addressed. The process of combining a number of different simulators together in this way presents some very difficult implementation and signal mapping problems.

3) **Fully integrated approach (I):**   This is the most flexible and most efficient approach of the three mentioned here. In this case, the various simulation algorithms are tightly-coupled and conform to a set of policies defined within the simulator for time-advancement and backup, signal mapping, etc. The algorithms are usually tailored for the mixed-mode environment and can handle bidirectional elements in a consistent manner. In addition, designers make use of a uniform interface to the simulator. The drawbacks of this approach are the long development time for this new program and the support and maintenance associated

with it.

Table 1.2 contains a partial list of mixed-mode simulators that have been reported in the literature. Note that they vary widely in the simulation modes that are supported. Although a detailed summary of the algorithms and techniques is not provided here, the reader is encouraged to consult the references for each program name to obtain further details. The mixed-mode simulators that combine process, device, and circuit level simulations have not been included in the table as they are beyond the scope of this book.

1.4. OUTLINE OF THE BOOK

This book focuses on the implementation of fully-integrated mixed-mode simulation and describes event-driven, relaxation-based techniques used in the SPLICE family of programs. While the issues of combining gate, RTL and behavioral levels of simulation are important, they often reduce to simple implementation issues. This book addresses the problem of mixing electrical simulation with gate-level simulation. Since electrical simulation is continuous in nature whereas gate-level simulation is discrete in nature, this particular problem presents a much more interesting challenge.

In Chapter 2, the electrical simulation problem is formulated, and the standard numerical techniques used to solve the problem are presented. Next, the issues associated with the implementation of an efficient time-step control scheme are described. This includes a description of the constraints imposed on the step size by the numerical methods; this is followed by two common time-step control schemes used in circuit simulation programs. In Chapter 3, two properties of waveforms, called *latency* and *multirate behavior*, are defined and used to motivate the need for new simulation methods. Then, the relaxation

PROGRAM	Type	B	R	G	S	T	E
ADLIB-SABLE[HIL80]	I	X	X				
ANDI(Silvar-Lisco)	I	X		X			X
DIANA[DEM81A]	I			X		X	X
DECSIM/SPICE[GRE88]	G	X	X	X			X
FIDELDO[TAH87]	G				X		X
LSIM/HSPICE (Silicon Compiler)	G	X	X	X	X	X	X
MOTIS3[CHE84B]	I	X			X	X	
PSPICE(MicroSim)	I	X		X			X
SALT(CAD Group)	I			X			X
SAMSON[SAK81]	I			X			X
SAMSON2[BEA86]	I			X			X
SABER/CADAT (Analogy/HHB)	G	X		X			X
SISYPHUS[GRO87]	I	X		X	X		X
SPLICE[NEW78B]	I			X		X	X
SPLICE1[SAL83]	I			X	X		X
SPLICE2[KLE84]	I		X	X		X	X
SWAN[DUM86]	I				X	X	X
VIEWSIM/AD[COR88]	G			X			X
iSPLICE3[ACU89]	I			X		X	X
iDSIM[OVE88]	I					X	X

Table 1.2: Survey of Mixed-Mode Simulators and Their Capabilities

methods are introduced and their convergence properties are described. First, the linear Gauss-Jacobi (GJ) and Gauss-Seidel (GS) methods are reviewed.   Then, the nonlinear relaxation and waveform relaxation methods are described. The requirement for partitioning to improve the convergence speed of relaxation methods is presented at the end of the chapter.

In Chapter 4, a number of algorithms based on nonlinear relaxation methods are described. A technique which combines nonlinear relaxation [ORT70] with *event-driven, selective-trace* [SZY75] to exploit waveform latency is presented. This approach is referred to as Iterated Timing Analysis or ITA [SAL83]. Its name is derived from the original work on "timing" simulation pioneered in the MOTIS program [CHA75]. The details of the implementation of ITA are provided.

Gate simulation is addressed in Chapter 5 and switch-level simulation is described in Chapter 6. Chapter 5 begins with a description of the evolution of logic state models and delay modeling. The Elogic technique for switch-level timing simulation and modeling is presented in Chapter 6. To conclude this chapter, the use of the Elogic modeling approach to resolve the signal mapping problems at the interface between electrical and logic elements is described.

In Chapter 7, the implementation details of the SPLICE mixed-mode simulator are presented. First, the implementation of event schedulers and event-driven, selective-trace techniques is detailed. Then, the overall architectural issues are described; this is followed by a summary of the transient analysis techniques used and event scheduling policies enforced between the different levels of simulation. Techniques for the dc solution of mixed-mode circuits are outlined, and mixed-mode simulation examples are provided to close out the chapter.

A summary, a number of directions for future work, and final conclusions are provided in Chapter 8.

CHAPTER 2

ELECTRICAL SIMULATION TECHNIQUES

The features of circuit or electrical simulation are extremely important in mixed-mode simulation as they determine the overall speedup and efficiency of the simulator. This chapter describes the basic theory and foundations for the electrical simulation techniques. First the circuit equations are formulated in Section 2.1 and the standard techniques are described in Section 2.2. The issues pertaining to time step selection and simulation accuracy are also addressed. The limitations of these techniques for large problems are identified and alternative approaches are described in the next chapter.

2.1. EQUATION FORMULATION

General-purpose circuit simulation programs such as ASTAP [WEE73], SPICE2 [NAG75] and SLATE [YAN80] provide a variety of analysis types including dc analysis, time-domain transient analysis, ac analysis, noise analysis and distortion analysis. By far the most CPU-intensive of these analyses is the time-domain transient analysis. The transient analysis problem involves computing the solution of a system of coupled nonlinear differential-algebraic equations over some interval of time, [0,T]. The most general form for the equations describing the circuit behavior is

$$F(\dot{x}(t), x(t), u(t)) = 0 \quad x(0)=X \quad (2.1)$$

where, $x(t) \in \mathbb{R}^n$ is the vector of unknowns, and may be a mixture of node voltages, branch currents, capacitive charges or inductive fluxes, $u(t) \in \mathbb{R}^r$ is a vector of independent sources, $F: \mathbb{R}^n \times \mathbb{R}^n \times \mathbb{R}^r \to \mathbb{R}^n$, and the initial conditions, $x(0)$, are specified by the vector X.

Equations of this form arise as a result of the properties of general electronic circuits. For example, the current through a capacitor is a function of the time derivative of the voltage across the capacitor; therefore, Eq. (2.1) is dependent on $\dot{x}(t)$. Since many devices have nonlinear relationships between their currents and voltages, **F** is also usually nonlinear. And finally, as a circuit is constructed from a collection of sparsely connected elements, **F** is a sparse function of the components of **x**. These circuit properties all have some impact on the numerical techniques used to solve the transient simulation problem and the resulting efficiency with which the solution is obtained.

There are a number of different ways to formulate the circuit equations described by Eq. (2.1). The most popular of these are Nodal Analysis (NA) [DES69], Modified Nodal Analysis (MNA) [HO75] and Sparse Tableau Analysis (STA) [HAC71]. These formulations are all based on the application of Kirchoff's Current Law (KCL), Kirchoff's Voltage Law (KVL) and the branch constitutive equations [DES69]. Nodal Analysis is the simplest of the three approaches. It uses KCL, which requires that the sum of the currents entering each node equals the sum of the currents leaving each node. In a circuit containing n+1 nodes, if KCL is written for every node in the circuit, a system of n equations is obtained assuming that one node is defined as a reference node. The currents in each equation can be replaced with the branch constitutive relations which are functions of the branch voltages (by assumption in NA), and KVL can be used to replace the branch voltages by node voltages. KVL requires that the sum of the voltages around any loop in a circuit be identically zero. The n node voltages are the unknown variables in this formulation. Note that it must be possible to represent the element and input source currents in terms of their terminal voltages in order apply Nodal Analysis. This requirement excludes current-controlled devices, floating voltage sources[1] and inductors and, therefore, limits the scope of the NA technique. However, inductors and floating voltage sources can be included in NA by simply reorganizing their branch equations as described in [MCC88, WHI85C]. Since the other current-controlled devices are not frequently used in the simulation of integrated circuits, NA is an adequate formulation technique for most practical circuits.

The formulation used throughout the rest of this book is Nodal Analysis. The NA equations are formulated as follows: First, KCL is applied at each node in a circuit with n nodes and b branches to produce a matrix equation of the form:

$$\mathbf{A} \, \mathbf{i} = 0 \qquad\qquad (2.2)$$

where $\mathbf{A} \in \mathbb{R}^{n \times b}$ is the reduced incidence matrix with entries of either +1, -1 or 0 and $\mathbf{i} \in \mathbb{R}^{b}$ is the vector of branch currents in the circuit. Element a_{ik} of $\mathbf{A}$ is +1 if a particular branch current, i_k, leaves node i, -1 if it enters node i and 0 if it is not incident at node i. If the set of branch currents is divided into the capacitor currents, i_c, and the currents through the resistive elements, i_r, then Eq. (2.2) can be rewritten as

$$\mathbf{A}_c \mathbf{i}_c = - \mathbf{A}_r \mathbf{i}_r \qquad\qquad (2.3)$$

where $\mathbf{A} = [\mathbf{A}_c \,|\, \mathbf{A}_r]$ and $\mathbf{i} = [i_c, i_r]^T$.

Each of the currents due to the nonlinear resistive elements can be replaced by their branch constitutive relations which are all functions of the branch voltages by assumption. The branch voltages, v_b, can be replaced by the node-to-datum voltages, v, using the relation:

---

[1] These are voltage sources with neither terminal connected to the ground node.

$$A^T v = v_b \qquad (2.4)$$

which follows from KVL [CHU75]. Then, the right-hand side of (2.3) can be written as

$$A_r i_r = - \begin{bmatrix} f_1(v) \\ f_2(v) \\ . \\ . \\ f_n(v) \end{bmatrix} \qquad (2.5)$$

where $f_k(v)$ is the sum of all the currents through the resistive elements connected to node k as a function of the node voltages, v.

The left-hand side of Eq. (2.3) represents the capacitor currents. The nonlinear capacitors are often specified in terms of their stored charge, q, a function of the voltage across the capacitor, v_c, as follows:

$$q = q(v_c)$$

The current flowing through the capacitor can be obtained by taking the time derivative of charge, which can then be related to the capacitance by applying the chain rule:

$$i_{cap} = \dot{q}(v_c) = \frac{dq(v_c)}{dv_c} \frac{dv_c}{dt} = C(v_c)\dot{v}_c \qquad (2.6)$$

Hence, each of the components of i_c in Eq. (2.3) can be replaced by $C(v_c)\dot{v}_c$. If Eq. (2.4) is used to replace the branch voltages by node voltages, then $A_c i_c$ can be transformed into the following:

$$A_c i_c = \begin{bmatrix} C_{11}(v) & . & . & C_{1n}(v) \\ . & . & . & . \\ . & . & . & . \\ C_{n1}(v) & . & . & C_{nn}(v) \end{bmatrix} \begin{bmatrix} \dot{v}_1 \\ . \\ . \\ \dot{v}_n \end{bmatrix} \qquad (2.7)$$

An important assumption which is sufficient to guarantee convergence of relaxation-based simulation techniques (to be described shortly) is that a two-terminal capacitor exists between each node and the reference node. These are referred to as *grounded capacitors*. This requirement is easily satisfied in real circuits where lumped capacitances are always present between circuit nodes and ground in the form of interconnect capacitance, and also between the terminals of active circuit elements and ground in the form of parasitic capacitances. Each grounded capacitor contributes a term to the diagonal of the capacitance matrix. Therefore, the C_{ii} elements are non-zero for all i. Note that C_{ij} is zero only if a capacitor does not exist between nodes i and j in the circuit.

By combining Eqs. (2.5) and (2.7), one obtains:

$$
\begin{bmatrix} C_{11}(v) & \cdots & C_{1n}(v) \\ \cdot & \cdot\ \cdot & \cdot \\ \cdot & \cdot\ \cdot & \cdot \\ C_{n1}(v) & \cdots & C_{nn}(v) \end{bmatrix} \begin{bmatrix} \dot{v}_1 \\ \cdot \\ \cdot \\ \dot{v}_n \end{bmatrix} = - \begin{bmatrix} f_1(v) \\ \cdot \\ \cdot \\ f_n(v) \end{bmatrix} \qquad (2.8)
$$

This equation can be written in the compact form:

$$
C(v(t),u(t))\ \dot{v}(t) = - f(v(t),u(t)), \quad t \in [0,T] \qquad (2.9)
$$

$$
v(0) = V
$$

where $v(t) \in \mathbb{R}^n$ is the vector of node voltages at time t, $\dot{v}(t) \in \mathbb{R}^n$ is the vector of time derivatives of $v(t)$, $u(t) \in \mathbb{R}^r$ is the input vector at time t, $C(x(t),u(t))$ represents the nodal capacitance matrix, and

$$
f(v(t),u(t)) = [f_1(v(t),u(t)), \cdots, f_n(v(t),u(t))]^T
$$

where $f_k(v(t),u(t))$ is the sum of the currents charging the capacitors connected to node k.

Equation (2.9) is a set of coupled first-order nonlinear differential equations that uses voltage as a state variable. This is commonly referred to as the capacitance formulation of the transient analysis problem. Alternatively, charge may be used as a state variable rather than voltage. The proper choice of voltage or charge as the state variable depends on the nature of the capacitors in the circuit. If all capacitances are linear, then either voltage or charge may be used as the state variable. However, in circuits with nonlinear capacitors, such as MOS circuits, charge must be used as the state variable due to considerations of charge conservation. That is, in order to keep the total charge in the system constant during the simulation process, charge must be used as the state variable. Examples of charge conservation problems arising from the use of Eq. (2.9) are given in [WAR78, YAN83, WHI85C].

The charge formulation of the circuit equations in normal form is given by

$$\dot{q}(t) = i(q(t))$$

where $q_k(v)$ is the sum of the charges due to the capacitors connected to node k and $i_k(q)$ is the sum of the currents charging the capacitors at node k. This equation can be solved to obtain the node charges as a function of time. However, information about charge is of little interest to the circuit designer, who would prefer to have information about the node voltages from the simulator. Therefore, it is preferable to write the charge formulation as

$$\dot{q}(t) = i(q(t)) = -f(v(t))$$

which is obtained by combining Eq. (2.6) and Eq. (2.9). This assumes that q is an invertible function of v. The charge formulation, including the input sources, $u(t)$, is given by

$$\dot{q}(v(t),u(t)) = -f(v(t),u(t)) \qquad (2.10)$$

Both the formulations given by Eqs. (2.9) and (2.10) will be used throughout this book.

2.2. STANDARD TECHNIQUES FOR TRANSIENT ANALYSIS

Equations (2.9) and (2.10) formulated above for the transient analysis of circuits must be solved using numerical techniques since, in general, it is difficult to obtain closed-form solutions. The first step is to apply a numerical integration method to discretize the time derivative, $\dot{x}(t)$. An integration method divides the continuous interval of time, $[0,T]$, into a set of M discrete time points defined by

$$t_0 = 0 , \quad t_{n+1} = t_n + h_n , \quad t_M = T. \tag{2.11}$$

An algebraic problem is solved at each time point, t_{n+1}, to obtain a sequence approximation to the exact solution. The quantity h_n is referred to as a time step. The selection of proper time-steps for a given problem is an important issue which is described in detail in Section 2.3. An example of a first-order implicit integration method is the backward-Euler (BE) method. To solve $\dot{x}(t)=f(x(t))$ using BE, the following expression is used:

$$x(t_{n+1}) = x(t_n) + h_n f(x(t_{n+1})) \tag{2.12}$$

This equation is implicit in that $x(t_{n+1})$ appears on both sides of the equation.

A numerical integration method converts a set of nonlinear differential equations into a set of nonlinear algebraic equations. These algebraic equations must be solved using some numerical method at each time point. The most commonly used method to solve nonlinear equations is the Newton-Raphson method [ORT70]. To solve a system of nonlinear equations, given by $F(x)=0$, using the Newton-Raphson method, the following iterative equation is used:

$$J_F(x^k)(x^{k+1} - x^k) = -F(x^k) \qquad (2.13)$$

where $J_F(x)$ is the Jacobian matrix and k is the iteration counter for the method. Each term in the Jacobian matrix, g_{ij}, is given by

$$g_{ij} = \frac{\partial F_i}{\partial x_j} \qquad (2.14)$$

where F_i is the ith component of F and x_j is the jth component of x. Equation (2.13) is iterated until $||x^{k+1} - x^k|| < \varepsilon_1$ and $||F(x^{k+1})|| < \varepsilon_2$. Note that if the problem is linear, then the Newton method produces the correct solution in one iteration.

The Newton method described above converts the set of coupled nonlinear algebraic equations into a set of coupled linear equations given by $Ax = b$, where $x \in \mathbb{R}^n$, $b \in \mathbb{R}^n$, $A \in \mathbb{R}^{n \times n}$ and A is assumed to be nonsingular. The matrix A is relatively sparse, typically having three elements per row [NEW83]. There are essentially two approaches to solving a sparse linear system. One approach is to use *direct methods* (such as LU decomposition) which attempt to exploit the sparse nature of the matrix during the computation. The implementation of these methods involves carefully chosing a data structure and the use of special pivoting strategies to minimize fillins [KUN86]. A second approach to the sparse linear problem is to use *relaxation methods*. The relaxation process involves decoupling the system of equations and solving each equation separately. An iterative method is applied between the equations until convergence is obtained. In effect, the problem of solving one large system containing n variables is converted to the problem of solving n subsystems each containing one variable.

The standard approach to circuit simulation is based on direct methods and uses the following steps:

1. MNA is used to formulate the system of differential-algebraic equations for the circuit.

2. Implicit integration methods are applied to convert the differential equations into a sequence of algebraic equations, which are nonlinear in general.

3. A damped Newton-Raphson method is used to convert the nonlinear equations into linear equations.

4. Direct sparse-matrix techniques are used to solve the linear equations generated by the Newton-Raphson method.

The details of the implementation of this approach in SPICE2 may be found in [NAG75]. This approach has proven to be very reliable and can be used across a variety of different technologies and element types. The most computationally intensive part of this approach is the Newton-Raphson iteration. It is composed of two phases: the formulation phase and the solution. These two phases, represented by steps 3 and 4 above, are repeated at each time point until convergence is obtained. In the formulation phase, the elements in the circuit are processed by calculating their contribution to the Jacobian matrix and the right-hand side vector in Eq. (2.13) to form the system of linear equations. This is also referred to as the function evaluation (or model evaluation) and load phase, and can be very time-consuming because of the complexity of the equations describing the elements in the circuit. For small to medium sized circuits containing MOS devices, the model evaluation and load times dominate the total CPU-time for the simulation [NEW78A].

In the second phase of the Newton iteration, the linear equations generated in the first phase are solved using direct methods such as LU decomposition. While this portion has a negligible contribution to the total run time for small circuits, it can in fact dominate the run time for very large circuits (i.e., greater than 1000 nodes in the circuit for SPICE2) [NEW83], as shown in Fig. 2.1. Therefore, any technique which attempts to reduce overall circuit simulation run times must reduce both the model evaluation time and the linear equation solution time to be effective.

2.3. TIME-STEP CONTROL: THEORETICAL ISSUES

Time-step control is an important issue in electrical simulation. In this section, the constraints imposed by the numerical techniques on the step sizes used in the integration process are described. Based on these constraints, an efficient time-step control scheme is presented. Ways to further improve the efficiency by using different step sizes to solve different components in the system are presented in the next chapter.

The circuit simulation problem, in its most general form, involves the solution of a system of nonlinear algebraic-differential equations. To simplify the description to follow, the circuit equations are assumed to be a system of differential equations in *normal form*:

$$\dot{x}(t) = f(x(t),u(t)) , \quad x(0) = X , \quad t \in [0,T] \tag{2.15}$$

where u is the set of primary inputs, x is a vector of unknown circuit variables and f is some nonlinear function. The vector of values specified as X are the initial conditions, and the simulation interval is $[0,T]$.

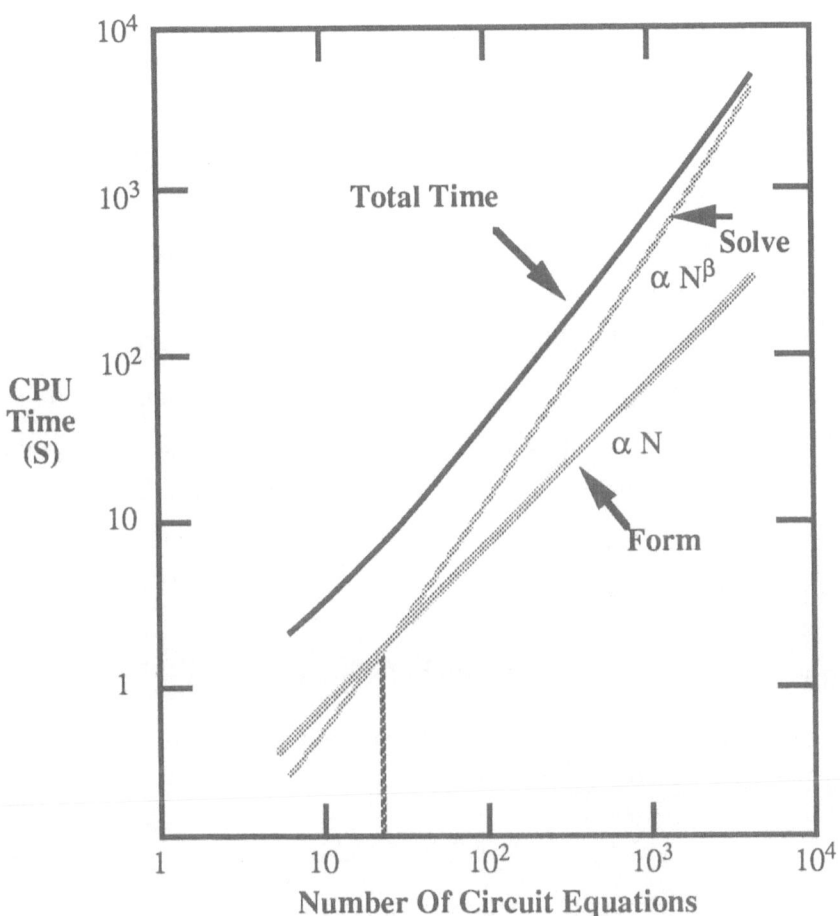

Figure 2.1:   CPU-Time vs. Circuit Size in SPICE2

2.3.1. Constraints on Step Size

The general form of a kth-order linear multistep integration method [GEA71] is given by

$$x_{n+1} = \sum_{i=0}^{p} a_i x_{n-i} + \sum_{i=-1}^{p} h_{n-i-1} b_i \dot{x}_{n-i} \qquad (2.16)$$

where x_n is the computed solution at time t_n, and h_n is the time-step at time t_n. The $2p+3$ coefficients, a_i and b_i, are chosen such that Eq. (2.16) will give the exact solution if the true solution is a polynomial in t of degree less than or equal to k [CHU75].

There are two broad classes of integration methods: explicit and implicit[2] [CHU75]. Explicit methods use only the solutions at previous time points to generate the solution at the next time point, and are characterized by $b_{-1}=0$. A number of explicit integration methods can be derived directly from a Taylor series expansion of $x(t)$ at the point t_n:

$$x_{n+1} = x_n + h_n \dot{x}_n + \frac{h_n^2}{2} \ddot{x}_n + \cdots \qquad (2.17)$$

For example, the forward-Euler (FE) method is obtained by taking the first two terms of Eq. (2.17):

$$x_{n+1} = x_n + h_n \dot{x}_n \qquad (2.18)$$

This *difference* equation can be formulated in terms of Eq. (2.17) by setting $p=0$, $a_0=1$, $b_0=1$ and all other coefficients to zero. Equation (2.18) implies that each equation can be updated independently, and in parallel, at each time point. For differential equations in the normal form, the solution at each time point can be obtained in one step and does not involve a matrix solution; therefore, the explicit methods are

---

[2] Recently, a number of combined integration-relaxation methods used in Timing Simulation [CHA75] have been classified as semi-implicit integration methods [DEM80, NEW83, WHI85C].

extremely efficient. Unfortunately, these methods are not as useful as implicit methods for circuit simulation. Implicit methods are characterized by $b_{-1} \neq 0$ in Eq. (2.16). The backward-Euler (BE) implicit integration method can be derived using a Taylor expansion of $\dot{x}(t)$ about the point t_n:

$$\dot{x}_{n+1} = \dot{x}_n + h_n \ddot{x}_n + \frac{h_n^2}{2} \frac{d^3 x_n}{dt^3} + \cdots \qquad (2.19)$$

Using Eq. (2.19) to replace $\dot{x}_n$ in Eq. (2.17), and ignoring the higher-order terms, the BE scheme is obtained:

$$x_{n+1} = x_n + h_n \dot{x}_{n+1} \qquad (2.20)$$

In this case, $p=0$, $a_0=1$, $b_{-1}=1$ with all other coefficients equal to zero. For nonlinear problems, this implicit equation is usually solved using an iterative method, often requiring a matrix solution. Therefore, the implicit methods are computationally more expensive than explicit methods. The forward-Euler and backward-Euler methods are representative of their respective class of integration algorithms and will be used to illustrate a number of other properties below.

a. Accuracy Constraint

Integration methods provide a numerical approximation to the true solution since, in general, the exact solution of Eq. (2.15) cannot be obtained. The error in the numerical solution is due to a combination of the machine error and the truncation error. The machine error is usually in the form of a round-off error, since finite precision arithmetic is used, and it depends on the floating-point arithmetic unit of the computer being used. The truncation error results from the fact that the Taylor series is truncated after a number of terms and this error depends on the specific integration method. The *local truncation error* (LTE) for general multistep methods is defined as

$$LTE_{n+1} = x(t_{n+1}) - x_{n+1} \tag{2.21}$$

where $x(t_{n+1})$ is the exact solution to Eq. (2.15) at t_{n+1}, and x_{n+1} is the computed solution obtained from Eq. (2.16). In this definition, it is assumed that $x(t_n)=x_n$ and, therefore, only provides information about the error which occurs over a single time-step, hence, its name "local" truncation error. The LTE for the forward-Euler method can be derived using Eq. (2.18):

$$LTE_{n+1} = x(t_{n+1}) - x_n - h_n \dot{x}(t_n) \tag{2.22}$$

Using a Taylor expansion for the first term about t_n, the LTE is given by the first remainder term of the resulting expression:

$$LTE_{n+1} = \frac{h_n^2}{2} \ddot{x}(\xi) \qquad t_n \le \xi \le t_{n+1} \tag{2.23}$$

If E_A is some user allowable error tolerance for the problem, the accuracy constraint is

$$\frac{h_n^2}{2} \ddot{x}(\xi) \le E_A \qquad t_n \le \xi \le t_{n+1} \tag{2.24}$$

This presents a bound on the step size which is given by

$$h_n \le \sqrt{2E_A / \ddot{x}(\xi)} \tag{2.25}$$

If this constraint is not satisfied, the solution must be rejected and a new solution computed with a smaller step size. Since the exact value of ξ is not known, the LTE is usually estimated using techniques to be described in a section to follow.

The backward-Euler method has an LTE given by

$$LTE_{n+1} = x(t_{n+1}) - x_n - h_n\dot{x}(t_{n+1}) \tag{2.26}$$

By expanding $x(t_n)$ in a Taylor series about t_{n+1} and applying the results to Eq. (2.26), the LTE is obtained by retaining the first remainder term:

$$LTE_{n+1} = -\frac{h_n^2}{2}\ddot{x}(\xi) \qquad t_n \leq \xi \leq t_{n+1} \tag{2.27}$$

Note that the error made in one step is $O(h^2)$ in both the FE and BE methods; hence, the accuracy bound on the step size is similar in both cases. However, the behavior of the global error, due to the accumulation of the local errors, may be quite different for the two methods and this difference strongly recommends the use of one method over the other. This characteristic is associated with the stability of the integration method.

b. Stability Constraint

The general stability characteristics of numerical integration methods applied to nonlinear differential equations are difficult to obtain. Usually the results are inferred from the analysis of a simple linear test problem [GEA71]:

$$\dot{x}(t) = -\lambda x(t) \ , \quad x(0) = x_0 \tag{2.28}$$

for which the solution is known to be

$$x(t) = x_0 e^{-\lambda t} \tag{2.29}$$

and, in general, λ is complex. This linear problem is useful because it is easy to analyze and provides information about the local behavior of nonlinear problems (i.e., when the step size is small). The problem is usually analyzed with $Re(\lambda) > 0$ so that the solution to Eq. (2.28) is stable. To further simplify the analysis, a fixed time-step is assumed. For example, if the FE method is used to solve Eq. (2.28), the following difference equation is obtained:

$$x_{n+1} = x_n - \lambda h x_n = x_n - \sigma x_n$$

where $\sigma = \lambda h$. Therefore,

$$x_{n+1} = (1 - \sigma)x_n = (1 - \sigma)^{n+1} x_0$$

The region of absolute stability is defined as the set of all complex values of σ such that x_{n+1} remains bounded as $n \to \infty$. For FE, it consists of all σ such that

$$|1 - \sigma| \leq 1 \qquad (2.30)$$

which produces the following constraint for real values of λ:

$$0 \leq \sigma \leq 2.$$

Therefore the time-step must lie in the range:

$$0 \leq h \leq \frac{2}{\lambda}. \qquad (2.31)$$

If step sizes outside this range are used, the computed solution will become unstable even though the true solution is stable. For BE, the difference equation is

$$x_{n+1} = x_n - \sigma x_{n+1}$$

Hence,

$$x_{n+1} = \frac{1}{(1+\sigma)^{n+1}} x_0$$

which results in the following requirement for stability:

$$\frac{1}{|1+\sigma|} \leq 1 \qquad (2.32)$$

Considering only real values of λ, the method produces a stable solution for all $h \geq 0$. Ideally, an integration method should produce a stable solution if the true solution is stable for any step size; this is the case for the BE method but not for FE. This property highly recommends the use of the BE method over the FE method since the step size can be selected based on accuracy considerations alone. For the general case when λ is complex, the region of Absolute stability for the BE integration method includes the entire right-half σ-plane. An integration method with this property is said to be A-stable [CHU75].

The forward-Euler and backward-Euler methods are examples of first-order integration methods. Higher-order methods with smaller local truncation errors can be constructed by taking more terms in the Taylor expansions of Eqs. (3.4) and (3.6). Integration methods with small LTEs are preferred as they allow larger time-steps to be used. For example, the trapezoidal method is a second-order integration method given by

$$x_{n+1} = x_n + \frac{h_n}{2}(\dot{x}_{n+1} + \dot{x}_n) \qquad (2.33)$$

and is quite popular as it is the most accurate A-stable method [CHU75]. The LTE for the trapezoidal method can be shown to be [CHU75]:

$$LTE_{n+1} = -\frac{h_n^3}{12}\frac{d^3x}{dt^3}(\xi) \qquad t_n \leq \xi \leq t_{n+1} \qquad (2.34)$$

Since the error is $O(h^3)$, it is often the case that a much larger step size can be used, compared to the BE method, for a given value of E_A.

c. Stiff-Stability Constraint

Another consideration in the choice of integration methods is the issue of stiffness. A stiff problem is one that exhibits time-scale variations of several orders of magnitude in the solution. A simple example of stiffness is the case of a fast initial "transient" in the solution, which dies quickly, followed by a slower "steady-state" solution. To handle this type of behavior, it is natural to use small time steps in the transient portion to accurately follow the solution and then to increase the step size for the remainder of the solution. However, this strategy may lead to instability of the integration method, especially for explicit integration methods. For example, if the test problem in Eq. (2.28) is solved using FE in the interval $[0,10^6\tau]$, where $\tau = 1/\lambda$, and $\lambda \in \mathbb{R}$, the time-step constraint given in Eq. (2.31) would be imposed in the entire interval even though the solution decays to zero in approximately 5τ. If the step size is increased beyond this stability bound, the solution will become unstable. On the other hand, if the size is kept within the constraint imposed by stability, the number of time points would be very large.

There are other situations which feature this kind of time-scale variation. A stiff problem is generated if the interval of time over which the system is integrated is large compared to the smallest time constant in the circuit, or if the circuit time constants themselves are widely separated. In addition, if the rise or fall time of an input waveform is widely separated from the circuit time constants, the problem also is considered to be stiff.

Integration methods which are appropriate for solving stiff problems should have regions of Absolute Stability which cover most of the right-half complex σ-plane so that the time-step can be selected based on the accuracy considerations alone. Explicit methods are not well-suited to stiff problems since their regions of Absolute Stability are usually very small. The A-stable integration methods are well-suited to stiff problems, but other implicit methods (for example, see [CHU75]) may be prone to instability when solving stiff problems. Gear proposed a family of integration methods called *stiffly-stable* methods [GEA71] which have the form:

$$\dot{x}_{n+1} = \frac{1}{h_n} \sum_{i=0}^{k} \alpha_i x_{n+1-i} \qquad (2.35)$$

The values for α_i are chosen such that a kth-order method is exact if the true solution is a kth-order polynomial. The methods of order k=1 and k=2 are both A-stable algorithms. The methods of order k=3 up to k=6 are not A-stable, but they do have stability regions which are quite suitable for the integration of stiff problems [GEA71]. These methods are also referred to as Backward-Differentiation Formulas (BDF) [BRA72]. A variable-order method, also proposed by Gear [GEA71], uses the integration order which allows the largest step size at each time point. This technique was implemented in the SPICE2 program [NAG75] and it was found that, even though the order could be varied from k=1 up to k=6, a second-order method was used most often in the computation. The reason for this was attributed to the nature of the nonlinearities in the circuit simulation problem (described in the next section) and nature of the solution waveforms. Therefore, most circuit simulators use a low-order implicit integration method with guaranteed stability properties so that the step sizes can be selected based on accuracy considerations alone.

2.3.2. Solution of Nonlinear Equations

When solving linear dynamic circuits, the accuracy and stability requirements of the numerical integration method are the only constraints on the step size used. Furthermore, linear problems can be solved in one "iteration" (i.e., one matrix solution) at each time point. Therefore, the amount of computation is directly proportional to the number of time points used. This is not true for nonlinear dynamic circuits, assuming that an implicit integration method is used. In fact, the cost of computing a solution at each time point is a function of the number of iterations used to solve the nonlinear algebraic problem. Consider the differential equation

$$\dot{x}(t) = f(x(t)) \tag{2.36}$$

where $f(x)$ is some nonlinear function. If the BE method is used to solve Eq. (2.36), the following equation is obtained:

$$x_{n+1} = x_n + hf(x_{n+1}) = G(x_{n+1}) \tag{2.37}$$

This nonlinear algebraic equation can be solved using a variety of techniques including fixed-point iteration and Newton's method. The approach usually taken in circuit simulators is to use Newton's method or one of its variants. Rewriting Eq. (2.37) as

$$F(x_{n+1}) = x_{n+1} - x_n - hf(x_{n+1}) = 0 \tag{2.38}$$

the Newton-Raphson method to solve this equation is given by the expression [ORT70]:

$$x_{n+1}^{k+1} = x_{n+1}^{k} - F(x_{n+1}^{k})/F'(x_{n+1}^{k}) \tag{2.39}$$

where k is the Newton iteration counter. In circuit terms, the Newton method replaces each nonlinear device in the circuit by a linearized model based on operating point information. This process converts the nonlinear circuit into a linear equivalent network. The linearized network is solved using standard linear circuit analysis techniques [CHU75]. The Newton method involves repeating the above steps until convergence is obtained.

To guarantee convergence of the Newton method, the functions $F(x)$ and $F'(x)$ must be continuous in an open neighborhood about x^*, $F'(x^*) \neq 0$, and the initial guess, x^0, must be close to the final solution. The Newton method is preferred over the simpler fixed-point method for several reasons. The main reason is that the fixed-point algorithm is not well-suited to stiff problems. It also imposes a bound on the time-step to guarantee convergence. Another reason is due to the quadratic convergence property of the Newton method. That is, if, in addition to the above conditions, $F''(x^*)$ exists, then for some $k>K$ the difference between successive iterations and the true solution satisfies the relation [ORT70]:

$$|x^{k+1} - x^*| \le c |x^k - x^*|^2$$

In practice, this quadratic convergence behavior occurs close to the final solution. Hence, it is important to provide an initial guess which is close to the final solution. In general, it is difficult to provide a reasonable starting guess for the Newton method. However, for the transient analysis problem it is possible to generate a good initial guess, especially if a capacitor exists between each node and the ground node[3]. For example, the solution at the previous time point is a good starting guess for the Newton method at t_{n+1}. A better approach is to use an explicit integration method [BRA72]:

$$x_{n+1}^0 = \sum_{i=1}^{k+1} \gamma_i x_{n+1-i} \qquad (2.40)$$

where the γ_i values are obtained by requiring that the predictor, x_{n+1}^0, be correct if the solution is a kth-order polynomial. Usually a kth-order predictor is used with a kth-order integration method.

The time-step also has some influence on the convergence speed of the Newton method. An intuitive reason for this can be given in circuit terms: the Newton method converts a nonlinear circuit into an associated linear circuit, as mentioned previously. As the step size is made smaller,

---

[3] A capacitor to ground at each node implies some smoothness in the solution since it prevents instantaneous changes in the voltage at the node. Therefore, as $h \to 0$, $x_{n+1} \to x_n$.

the values of linearized circuit elements begin to approach their values at the previous time point. Therefore, the circuit will behave almost linearly in this interval and convergence can be obtained in very few iterations, possibly even a single iteration. On the other hand, if the step size is too large, a good starting guess may be difficult to generate, and could lead to either slow convergence or nonconvergence. If nonconvergence should occur, the time-step must be rejected and a smaller step used in its place. Hence, in some cases, it may actually be more efficient to use two small steps rather than one large step.

2.4. TIME-STEP CONTROL: IMPLEMENTATION ISSUES

The simplest time-step selection scheme is to use the same time-step throughout the interval of interest, [0,T]. That is, use a *fixed* time-step. Unfortunately, there are a number of constraints on the step size which may require that h be extremely small, resulting in a large number of time points. These constraints arise from the accuracy, stability and stiff-stability properties of a numerical integration method. For a fixed-step approach, the step size would have be chosen such that it satisfies these constraints under worst-case conditions. A better approach is to vary the step size during the simulation in accordance with the variation in the constraints. For a given problem, the allowable step sizes depend primarily on the properties of the specific integration method being used. In this section, the main considerations in the implementation of an efficient time-step control for circuit simulation are described. It includes a discussion of LTE time-step control, iteration count time-step control and the effect of input sources on time-step selection.

2.4.1. LTE Time-Step Control

In LTE time-step control, the user provides two accuracy control parameters, ε_a and ε_r, which are the absolute and relative errors permissible in each integration step. They are combined to form a user error tolerance:

$$E_{UserLTE} = \varepsilon_a + \varepsilon_r \times max \, | \, x_{n+1}, x_n |$$

The general form of the local truncation error for most multistep integration methods of order **k** is given by [GEA71,CHU75]

$$LTE_{n+1} = \tilde{C}_k h^{k+1} x^{(k+1)}(\xi) \qquad t_n \leq \xi \leq t_{n+1} \qquad (2.41)$$

where $\tilde{C}_k$ is a constant which depends on the coefficients of Eq. (2.16) and the order of the method. Since the value of $x^{(k+1)}(\xi)$ is not known, in general, it must be estimated in some way using the numerical solutions. Typically a divided-difference approximation is used. The first divided-difference is defined as

$$DD_1(t_{n+1}) = \frac{x_{n+1} - x_n}{h_n}$$

and the k+1st divided-difference is defined as

$$DD_{k+1}(t_{n+1}) = \frac{DD_k(t_{n+1}) - DD_k(t_n)}{\sum\limits_{i=0}^{k-1} h_{n-i}}$$

Then the estimate for the derivative term in Eq. (2.41) is (see [NAG75] for derivation)

$$x^{(k+1)}(\xi) \approx (k+1)! \, DD_{k+1}(t_{n+1}).$$

The LTE estimate is then

$$E_k = C_k h^{k+1} DD_{k+1}(t_{n+1})$$

For the BDF integration methods [BRA72], the LTE can be estimated in a more convenient way. The estimate is the calculated using difference between the computed solution x_{n+1} and the predicted value $x^P(t_{n+1})$.

For a kth-order BDF method, the following expression is used:

$$E_k = \left[\frac{h_n}{t_{n+1} - t_{n-k}} \right] (x_{n+1} - x^P(t_{n+1}))$$

The expression for $x^P(t_{n+1})$ is given in Eq. (2.40). The computed solution x_{n+1} is accepted if

$$|E_k| < E_{UserLTE} \tag{2.42}$$

One way of implementing this check is to take the ratio of the allowable LTE and the actual LTE:

$$r = \frac{|E_{UserLTE}|}{|E_k|} = \frac{|C_k h_{max}^{k+1} \dddot{x}(\xi)|}{|C_k h_n^{k+1} \dddot{x}(\xi)|}$$

Noting that both errors are $O(h^{k+1})$, it follows that

$$r = \left[\frac{h_{max}}{h_n} \right]^{k+1}$$

and

$$r_{LTE} = \frac{h_{max}}{h_n} = (r)^{(\frac{1}{k+1})}$$

The comparison test given in Eq. (2.42) becomes

$$r_{LTE} > 1.0$$

to accept the computed solution. The advantage of this ratio is that it can also be used to select the step size for the next integration step. Therefore, the next recommended step size is given by

$$h_{n+1} = r_{LTE} h_n \tag{2.43}$$

In practice, Eq. (2.43) may occasionally recommend rather abrupt changes in the step size. A number of experiments have shown that rapid changes in step size may introduce stability problems [BRA72]. Intuitively, the step sizes should follow the smoothness of the solution. To ensure that the changes in the step size are indeed gradual, it is convenient to set upper and lower bounds on the changes in step size. In fact, three regions can be defined as follows:

- if $r_{LTE} < 1.0$, reduce the step size by MAX(s_l , r_{LTE})
- if $1.0 \leq r_{LTE} < \alpha$, maintain the same step size
- if $r_{LTE} \geq \alpha$, increase the step size MIN(s_u , $\beta\ r_{LTE}$)

The time step may be reduced at most by the factor s_l and increased at most by the factor s_u. The α factor permits the same step size to be used a number of times. Typically, $\alpha=1.2$, $s_l=0.25$ and $s_u=2.0$. Note that a multiplying factor β has also been introduced as part of the growth factor. The β factor is a way of making the time step selection somewhat conservative. Since the LTE can only be estimated, it may occasionally be optimistic [YAN80]. If so, the time step would be rejected and a smaller step used unnecessarily. The β factor reduces the likelihood of this happening and a typical value is 0.9.

2.4.2. Iteration Count Time-Step Control

As mentioned before, the use of large steps is not necessarily the most efficient approach for nonlinear circuits, especially if relaxation is used. In fact, if the time step is too large, the iterative method may not converge, which would force the time step to be rejected, resulting in wasted effort. This suggests that the time step control should also be controlled by the nonlinearity of the problem.

A number of programs use a time step control based on nonlinearity considerations alone (e.g., SPICE2, ADVICE, MOTIS3) called *iteration count* time step control. This strategy minimizes the total number of Newton iterations used during the simulation. The step sizes are selected as follows. If the number of iterations is larger than N_{high}, the step size is reduced by some factor. If the number of iterations is less than N_{low}, the step size is increased by some factor. Otherwise, the step size remains the same. The idea is to use approximately the same number of iterations at each time point.

While this strategy is certainly effective at reducing the overall computation time, it is prone to accuracy problems [NAG75]. For example, for linear circuits the step size would always be increased since the solution is always obtained in one "iteration" at each time point. For weakly nonlinear circuits, the same sort of effect would be observed. Therefore, this approach, when used by itself, is not recommended since it does not control the numerical integration errors directly. However, the iteration count time-step control can be used in conjunction with the LTE-based time step control. In this case, if too many iterations were required to converge, a somewhat smaller step size could be used in the next integration step. If too few iterations are used, a slightly larger step size can be used. The method could be implemented by making the growth factor dependent on the number of iterations used to compute the solution. Of course, if convergence is not obtained in a specified number of iterations, the time step should be rejected and a smaller step used in its place.

CHAPTER 3

RELAXATION-BASED SIMULATION TECHNIQUES

The overall goal in circuit simulation is to generate the solution as efficiently as possible while providing the desired level of accuracy. As described in the last chapter, the standard approach to solving Eq. (2.1) is to use a numerical integration method. One way to make the integration process efficient is to simply minimize the total number of time points used. That is, at any stage during the simulation, take the largest step possible that provides the required accuracy. This strategy is effective for linear problems, assuming that the numerical integration method has guaranteed stability properties but does not guarantee a smaller runtime. In fact, for nonlinear problems, it may be more efficient to take smaller steps so that the iterative method used to solve the nonlinear algebraic equations converges in fewer iterations. Using small time-steps also improves the accuracy of the solution. Therefore, minimizing the total number of iterations is a more useful goal in reducing the amount of computation.

The cost of each iteration is proportional to the number of model evaluations[1] performed, assuming that the linear equation solution time is small. Therefore, the number of model evaluations used in the solution process is a good measure of the amount of computation. Based on this argument, the objective for the efficient solution of the differential equations in Eq. (2.1) should be to minimize the total number of model evaluations.

---

[1] A model evaluation usually refers to the calculation of the currents and conductances for a MOS or bipolar transistor, or some equivalent amount of computation.

A number of researchers have attempted to reduce the computation time for expensive model evaluations by using lookup tables for active devices [CHA75, NEW79, BUR83, GYU85]. In this approach, a number of tables of device characteristics are generated prior to the analysis, and simple table lookup operations are performed during the analysis in place of the expensive analytic evaluations. Points which are not available in the tables are interpolated using polynomial interpolation or splines. One drawback of this approach is that there may be a substantial memory requirement for these tables, depending on the level of accuracy desired, but it is usually justified by the improvement in computation speed. Current research in this area involves reducing the memory requirements without sacrificing either the computational advantage or the accuracy of the device models. Further details on this topic may be found in the references listed above.

In this chapter, the focus is on reducing the total number of expensive model evaluations by minimizing the number of time points computed for each waveform. This is accomplished by using relaxation-based techniques to exploit the waveform properties such as latency and multirate behavior. Section 3.1 begins by introducing the general concepts of waveform latency and multirate behavior. In Section 3.2, the various relaxation-based techniques that are used to exploit latency and multirate behavior are examined. In Section 3.3, the circuit partitioning issues for relaxation methods are addressed.

3.1. LATENCY AND MULTIRATE BEHAVIOR

Most circuit simulators employing direct methods use a single common time step for the whole system and, hence, compute the solution of every variable at every time point. The time-step at each point is based on the fastest changing variable in the system, i.e., the $n+1$st time point is given by

$$t_{n+1} = t_n + h_n$$

where h_n is the integration step size determined by

$$h_n = \min(h_{1,n}, h_{2,n}, \cdots, h_{N,n})$$

and $h_{i,n}$ is the recommended step size for with the ith variable at t_n. As a result, many variables are solved using time steps which are much smaller than necessary to compute their solutions accurately. For example, the computed points of a waveform from a large digital circuit, simulated using direct methods, are shown in Fig. 3.1(a). Note that there are many more points than necessary to represent the waveform accurately, especially in the regions when the waveform is not changing at all. The extra points are due to some other variable changing rapidly in the same region of time. The same waveform is shown Fig. 3.1(b) with only the minimum number of points necessary to represent it accurately.

Since the objective in circuit simulation is to provide an accurate solution while minimizing the number of expensive model evaluations, one way to achieve this goal is to reduce the number of time points computed for each waveform. A number of circuit simulators have attempted to improve the efficiency in this manner by exploiting a property of waveforms called *latency* [NAG75, NEW78, RAB79, YAN80, SAK81]. While the general concept of latency includes any situation where the value of a variable at a particular time point can be computed accurately using some explicit formula, it usually refers to the situation where a variable is not changing in time and its solution can be obtained from the explicit equation:

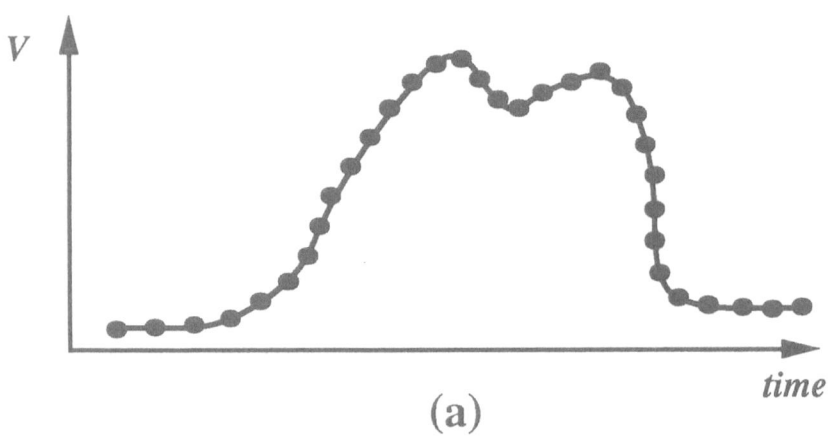

(a)

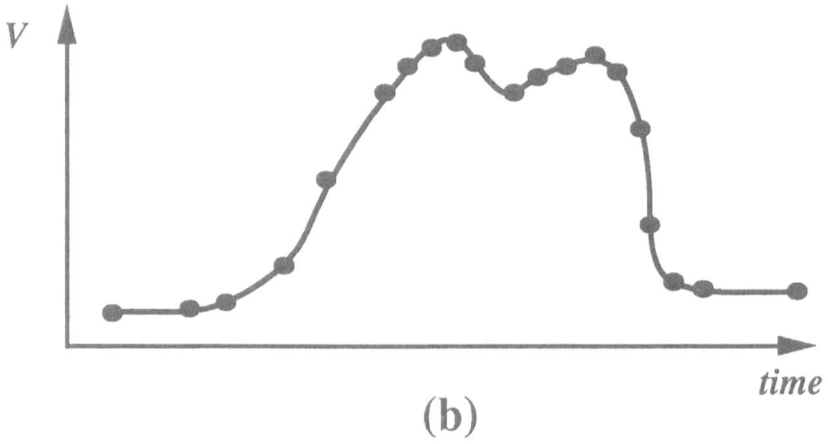

(b)

Figure 3.1: Effect of Solution by Direct Methods

$$x_{n+1} = x_n \qquad\qquad (3.1)$$

That is, the value x_{n+1} is not computed using a numerical integration formula but instead is simply updated using the value at the previous time point. For example, the waveform shown in Fig. 3.2(a) has three latent periods, and ideally the value of x does not need to be computed in any of these regions.

In the SPICE program [NAG75], latency exploitation is performed using a *bypass* scheme. In this technique, each device is checked to see if any of its associated currents and node voltages have changed significantly since the last iteration. If not, the same device conductances and current are also used in the next iteration. However, the checking operation is somewhat expensive, especially if the circuit is large and most of the devices are latent. In general, latency exploitation involves the use of a model describing the behavior of a particular variable as a function of time over a given interval. The simple model described in Eq. (3.1) can be considered as a "zeroth-order" latency model. Higher-order latency models can be constructed if the solution is known to have a specific form (i.e., polynomial, exponential) or if the solution for the variable can be obtained in closed form. For example, a first-order latency model given by

$$x_{n+1} = x_n + h_n \frac{I}{C}$$

can be used in the case of an ideal current source, with current I, charging a linear capacitor, C. Usually a latency model can only be used over a portion of the simulation interval. Therefore, the validity of the model must be monitored and its use must be discontinued when the model is thought to be invalid. The latency model used in this context has also been called a *dormant* model [SAK81].

In practice, only the zeroth-order form of latency can be exploited easily since the higher-order forms are difficult to construct for general nonlinear circuits. To exploit this simple form of latency, some mechanism is necessary to detect that the signal value is not changing appreciably[2]. The waveform is considered to be latent at that point, and its associated variable is updated using Eq. (3.1) at subsequent time points. A second mechanism is used to determine when the latency model is invalid, and from that point onward the variable is computed in the usual way. Hence, the waveform is only computed at time points when the signal is changing. Event-driven, selective trace can be used to exploit latency, as described in the next chapter, without incurring the overhead of the bypass scheme.

It is only useful to exploit this simple form of latency when some variables in the circuit are changing while other variables are stationary, since direct methods can adequately handle the case when all variables are active or latent. In fact, the "useful" form of zeroth-order latency can be viewed as a subset of a more general property of waveforms called *multirate behavior* which is illustrated in Fig. 3.2(b). Multirate behavior refers to signals changing at different rates, relative to one another, over a given interval of time. MOS circuits inherently exhibit this kind of behavior because of different transistor sizes and different capacitance values at each node. Exploiting this general property can reduce significantly the number of time points computed for each waveform since large steps can be used for variables changing very slowly while smaller steps can be used for rapidly changing variables.

The basic strategy to speed up circuit simulators suggested above is to take advantage of the relative inactivity of large circuits by reducing

---

[2] A number of schemes to detect zeroth-order latency are described in Chapter 4.

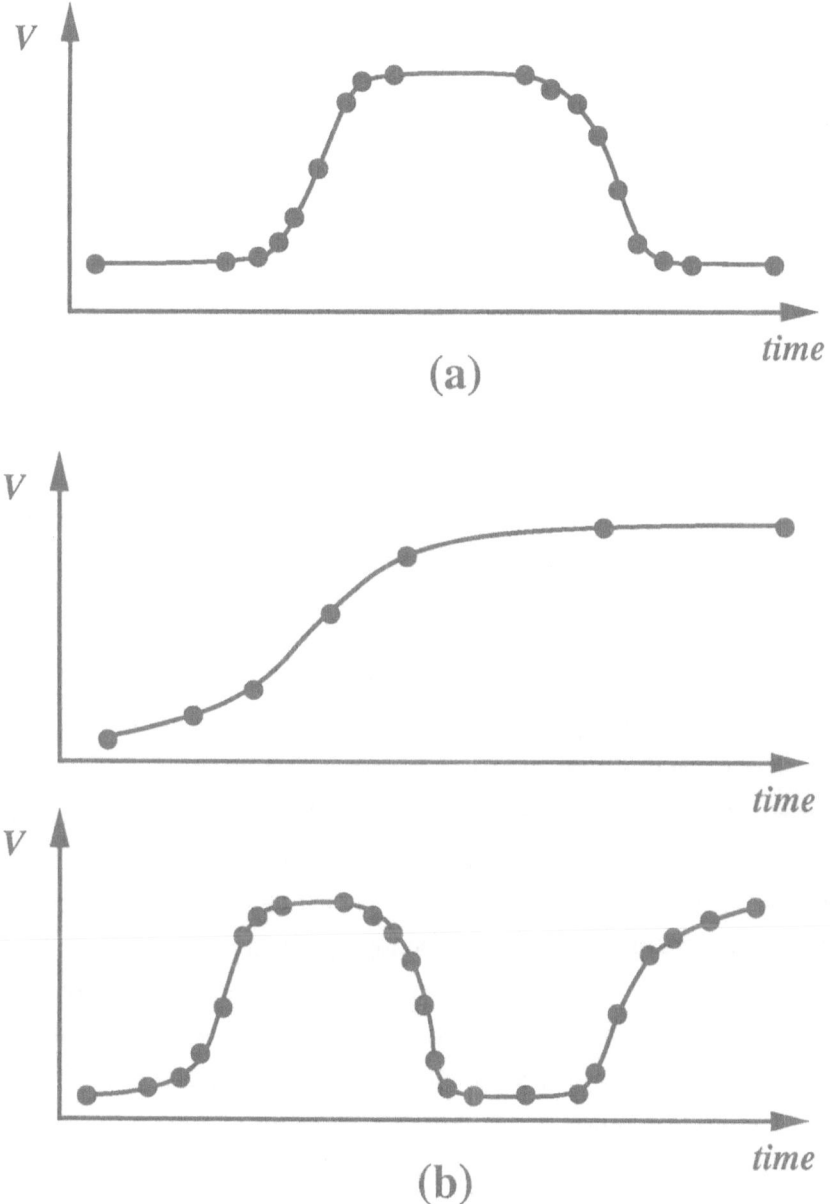

Figure 3.2: Waveform Properties
(a) Latency (b) Multirate Behavior

the number of time points computed. However, the actual speed improvement obtained by solving the equations in this manner depends on the two main factors:

    1) The "amount" of latency and multirate behavior exhibited
       by the circuit during the simulation, and

    2) The efficiency of techniques used to exploit the two properties.

The first point refers to the maximum speed improvement that can be obtained if the two waveform properties are exploited fully, and this factor depends on the circuit size and the activity in the circuit generated by the external inputs. The second factor depends on the actual number of points computed and the work required to compute each point.

3.2. OVERVIEW OF RELAXATION METHODS

Relaxation-based circuit simulators, such as SPLICE [SAL83, KLE84] and RELAX [LEL82, WHI83], use iterative methods at some stage of the solution process to solve the circuit equations. The success of these programs is due to the fact that they offer the same level of accuracy as direct methods, assuming identical device models, while significantly reducing the overall simulation run time. The reduction in run time is accomplished by computing fewer solution points for each waveform, thereby reducing the total number of model evaluations, and by avoiding the direct sparse-matrix solution. However, a tradeoff exists in the relaxation methods since they can only be applied to a specific class of circuits. Furthermore, there is the additional requirement that a grounded capacitor be present at each node in the circuit to guarantee convergence. While these factors limit the scope of the application of relaxation methods, the programs which use relaxation have proven to be extremely useful for simulation of many industrial MOS and bipolar integrated circuits. In the remainder of this chapter, the relaxation methods are described and their mathematical properties are presented.

3.2.1. Linear Relaxation

Two common linear iterative methods are the Gauss-Jacobi (GJ) and Gauss-Seidel (GS). The methods differ only in the information they use when solving a particular equation as shown in the two algorithms given below. The superscript k is the iteration count, and ε is some small error tolerance.

<u>Algorithm 3.1 (Gauss-Jacobi Method to solve $Ax = b$)</u>

    $k \leftarrow 0$;

    guess x^0 ;

    repeat {

        $k \leftarrow k+1$;

        forall ($i \in \{1, \cdots, n\}$)

$$x_i^k = \frac{1}{a_{ii}} \left[b_i - \sum_{j=1}^{i-1} a_{ij} x_j^{k-1} - \sum_{j=i+1}^{n} a_{ij} x_j^{k-1} \right] ;$$

    } **until** ($|x_i^k - x_i^{k-1}| \le \varepsilon$, $i=1, \cdots, n$);

■

<u>Algorithm 3.2 (Gauss-Seidel Method to solve $Ax = b$)</u>

    $k \leftarrow 0$;

    guess x^0 ;

    repeat {

        $k \leftarrow k+1$;

        foreach ($i \in \{1,..,n\}$)

$$x_i^k = \frac{1}{a_{ii}} \left[b_i - \sum_{j=1}^{i-1} a_{ij} x_j^{k} - \sum_{j=i+1}^{n} a_{ij} x_j^{k-1} \right] ;$$

    } **until** ($|x_i^k - x_i^{k-1}| \le \varepsilon$, $i=1, \cdots, n$);

■

Notice that in the GJ method each x_i^k is computed using the iteration values $x_j^{(k-1)}$, $j=1, \cdots ,n$, which are the values from the previous iteration. In the GS method, the latest iteration values are used as soon as they become available. The **forall** construct in Algorithm 3.1 suggests that all **n** variables can be computed in parallel during each iteration. The **foreach** construct in Algorithm 3.2 requires that the variables be processed in a particular sequence.

Linear relaxation schemes are usually described using a *splitting* notation that separates **A** into two components:

$$A = B - C$$

where **B** is a nonsingular matrix such that linear systems of the form **Bx = d** are "easy" to solve. Various relaxation schemes can be constructed by choosing different **B** and **C** matrices in the iterative equation:

$$x^{k+1} = - B^{-1}Cx^k + C^{-1}b$$

In particular, if **A** is decomposed into its diagonal, strictly lower-triangular and strictly upper-triangular parts, D, L and U, respectively such that $A = L + D + U$, then the GS method is obtained by setting

$$B = (L+D) \qquad C = - U \tag{3.2}$$

and the GJ method is obtained using

$$B = D \qquad C = - (L+U). \tag{3.3}$$

Since relaxation methods are iterative, the question naturally arises as to whether or not these methods converge to the correct solution and, if so, under what conditions? The requirements for convergence are stated in the following standard theorem [VAR62]:

Theorem 3.1: Suppose $\mathbf{b} \in \mathbb{R}^n$ and $A = B - C \in \mathbb{R}^{n \times n}$ is nonsingular. If $\mathbf{B}$ is nonsingular and the spectral radius of $\mathbf{B}^{-1}\mathbf{C}$, given by $\rho(\mathbf{B}^{-1}\mathbf{C})$, satisfies the condition $\rho(\mathbf{B}^{-1}\mathbf{C}) < 1$, then the iterates $\mathbf{x}^{(k)}$ defined by $\mathbf{B}\mathbf{x}^{(k+1)} = \mathbf{C}\mathbf{x}^{(k)} + \mathbf{b}$ converge to $\mathbf{x}^* = \mathbf{A}^{-1}\mathbf{b}$ for any starting vector $\mathbf{x}^{(0)}$.
■

In other words, the magnitude of the largest eigenvalue of the iteration matrix $\mathbf{B}^{-1}\mathbf{C}$ must be strictly less than 1 to guarantee convergence of a linear relaxation method. A condition which guarantees that $\rho(\mathbf{B}^{-1}\mathbf{C}) < 1$ is if $\mathbf{A}$ is strictly diagonally dominant. A matrix has this property if the diagonal term in each row i is greater than the sum of the off-diagonal terms in the same row, i.e.,

$$\sum_{\substack{j=1 \\ j \neq i}}^{n} |a_{ij}| < |a_{ii}| \quad \text{for } 1 \leq i \leq n$$

and the "more dominant" the diagonal, the more rapid will be the convergence. However, these linear relaxation methods have a linear convergence rate.

A number of techniques are available to improve the convergence speed of linear relaxation methods. For example, in the GS method, the order in which the equations are solved usually has a strong effect on the number of iterations required to converge. Consider the case when matrix A is lower triangular. If processed in the sequence, $x_1, x_2, \cdots, x_n$, then one relaxation iteration is sufficient to obtain the correct solution. However, if processed in the reverse order, then n iterations are required to obtain the solution. Therefore, equation ordering is usually performed on the variables whenever GS is used.

Another technique to improve convergence, also used in conjunction with the Gauss-Seidel method, is the method of Successive

Overrelaxation (SOR). In this approach, the Gauss-Seidel method is used initially to generate an intermediate value, $\tilde{x}_i^{(k+1)}$, using the equation

$$\tilde{x}_i^{(k+1)} = B^{-1}Cx_i^{(k)} + B^{-1}b$$

where B and C are defined by Eq. (3.2). The actual value of $x_i^{(k+1)}$ is obtained by taking a weighted combination of the previous iteration and the intermediate value which depends on a relaxation parameter, ω.

$$x_i^{(k+1)} = (1-\omega)x_i^{(k)} + \omega\tilde{x}_i^{(k+1)}$$

The SOR method can also be defined in terms of the splitting notation with $B = \omega^{-1}(D+\omega L)$, and $C = \omega^{-1}[(1-\omega)D - \omega U]$. While the proper choice of ω can greatly reduce the number of iterations, an optimal value of ω can only be computed *a priori* for a limited number of cases. In general, it may be necessary to perform a somewhat complicated eigen-value analysis to determine the best value of ω. In practice, adaptive algorithms are used to select an appropriate value for ω during the solution process.

Linear relaxation methods can be used in conjunction with the solution of nonlinear equations to solve the linear systems generated by Newton's method. For example, the Newton-SOR method is a combination of the Newton-Raphson method and the SOR method. In this composite algorithm, the Newton iteration can be considered as the "outer loop" and the SOR iteration as the "inner loop." While it is possible to carry the inner loop to convergence, there is no requirement to do so, as long as the outer loop is iterated to convergence. In general, an m-step Newton-SOR method can be defined where m is the number of iterations used in the inner loop. For the case $m=1$, a one-step Newton-SOR method is obtained. The Newton-SOR method is only one example of the possible combinations of nonlinear iterative methods and linear iterative methods. For example, Newton's method may be replaced by the secant method and the SOR iteration may be replaced by one of the standard Gauss-Seidel or Gauss-Jacobi methods.

3.2.2. Nonlinear Relaxation

The basic idea of relaxation can also be extended to solve systems of nonlinear equations of the form $F(x) = 0$, where $F:\mathbb{R}^n \to \mathbb{R}^n$, with components $f_1, f_2, \ldots, f_n$ and $f_i:\mathbb{R}^n \to \mathbb{R}$. That is, rather than solving the system using direct matrix techniques, the nonlinear equations can be solved in a decoupled fashion. Two such algorithms are given below. The index k is the iteration count, while ε_1 and ε_2 are error tolerances.

Algorithm 3.3 (Nonlinear Gauss-Jacobi Method to solve $F(x) = 0$)
$\qquad$ $k \leftarrow 0$; guess x^0 ;
$\qquad$ **repeat** {
$\qquad\qquad$ $k \leftarrow k+1$;
$\qquad\qquad$ **forall** ($i \in \{1, \cdots, n\}$)
$\qquad\qquad\qquad$ solve $\quad f_i(x_1^{k-1}, \cdots, x_{i-1}^{k-1}, x_i^k, x_{i+1}^{k-1}, \cdots, x_n^{k-1}) = 0$
$\qquad\qquad\qquad$ for x_i^k ;
$\qquad$ } **until** ($|x_i^k - x_i^{k-1}| \le \varepsilon_1$, $|f_i(x^{k,i})| \le \varepsilon_2$, $i=1, \cdots, n$);

∎

Algorithm 3.4 (Nonlinear SOR Method to solve $F(x) = 0$)
$\qquad$ $k \leftarrow 0$; guess x^0 ;
$\qquad$ **repeat** {
$\qquad\qquad$ $k \leftarrow k+1$;
$\qquad\qquad$ **foreach** ($i \in \{1, \cdots, n\}$)
$\qquad\qquad\qquad$ solve $f_i(x_1^k, \cdots, x_{i-1}^k, x_i^k, x_{i+1}^{k-1}, \cdots, x_n^{k-1}) = 0$ for
$\qquad\qquad\qquad$ x_i^k ;
$\qquad\qquad$ $x_i^k \leftarrow (1-\omega)x_i^{k-1} + \omega(x_i^k)$;
$\qquad$ } **until** ($|x_i^k - x_i^{k-1}| \le \varepsilon_1$, $|f_i(x^{k,i})| \le \varepsilon_2$, $i=1, \cdots, n$);

∎

In the above algorithms, $x^{k,i} = (x_1^k, \cdots, x_{i-1}^k, x_i^k, x_{i+1}^k, \cdots, x_n^k)$.

These algorithms are referred to as nonlinear relaxation methods. The steps are very similar to linear relaxation as given in Algorithms (3.2) and (3.3) except that, in this case, each equation in the inner loop is nonlinear. To solve each one-dimensional nonlinear problem, $f_i(x) = 0$, an iterative technique such as the Newton method or secant method must be used since, in general, a closed-form solution cannot be obtained. Combining the SOR method with the Newton method results in the SOR-Newton algorithm. The general case is the m-step SOR-Newton method, where m is the number of Newton iterations taken in the inner loop. The question again arises as to the number of inner loop iterations to use.

It can be shown that the rate of convergence of the one-step SOR-Newton method is the same as for the one-step Newton-SOR method [ORT70]. The m-step SOR-Newton method also has the same rate as the one-step method implying that it is not worthwhile to take more than one Newton step since the convergence rate is not affected. However, the convergence rate of the m-step Newton-SOR method is m times the rate of convergence of the one-step method. Therefore, based on the rates of convergence, one might be inclined to choose the m-step Newton-SOR to solve a system of nonlinear equations. There is, however, a hidden cost if the partial derivatives are expensive to calculate. Each step of SOR-Newton requires the evaluation of each f_i and n partial derivatives, $\dfrac{\partial f_i}{\partial x_i}$. whereas the m-step Newton-SOR method requires the evaluation of f and all partial derivatives. Based on both operation counts and the rates of convergence given above, the one-step SOR-Newton method appears to be the most efficient and for this reason it is used in Iterated Timing Analysis (ITA) [SAL83]. Note that this implies one iteration in the inner loop. The outer loop is iterated until convergence is obtained. SOR-Newton also offers one additional advantage over Newton-SOR in that waveform latency can be exploited easily. This feature is described in more detail in the chapter to follow.

In a general-purpose implementation of these methods, the iterative process must be terminated when the solution is close enough to x^*. Often, this condition is checked using the test $|x_i^k - x_i^{k-1}| \leq \varepsilon_1$. However, this check of convergence is not sufficient in the nonlinear case. A second test is necessary to ensure that each function, f_i, is close enough to zero, and this is specified using the test $|f_i(x^{k,i})| \leq \varepsilon_2$ for all i.

The algorithms presented above are meaningful only if the nonlinear equations, which are solved at each step in the inner loop, have unique solutions in some specific domain under consideration. Recall that for linear relaxation, the condition that $a_{ii} \neq 0$, for all $i=1, \cdots, n$ ensures that a solution exists, assuming that the diagonal dominance property holds. A similar condition is required in the nonlinear case. To illustrate this point, let the Jacobian be decomposed into its diagonal, strictly lower-triangular and strictly upper-triangular parts as follows:

$$F'(x) = D(x) + L(x) + U(x)$$

The iterations in the nonlinear scheme are well-defined if F is continuously differentiable in an open neighborhood S of the point x^*, for which $F(x^*)=0$, and $D(x^*)$ is nonsingular. The requirements for convergence are also analogous to those for the linear case. By splitting the Jacobian matrix using the previous notation

$$F'(x) = B(x) - C(x),$$

the local convergence of the nonlinear relaxation methods described in Algorithms (3.5) and (3.6) can be stated as follows [ORT70]:

Theorem 3.2: Given $F: \mathbb{R}^n \rightarrow \mathbb{R}^n$, assume that F is continuously differentiable in an open neighborhood S of x^* and x^* satisfies $F(x^*)=0$. If $B(x^*)$ is nonsingular and $\rho(B(x^*)^{-1}C(x^*))<1$, then there exists an open ball $S^* \subset S$ such that the nonlinear relaxation methods given in Algorithms (3.5) and (3.6) converge to x^* for any initial guess $x^0 \in S^*$. ∎

Recall that under the conditions stated in Theorem 3.1, linear relaxation methods converge for any initial guess. However, for the nonlinear case the convergence result is local since the initial guess must be close enough to the final solution to guarantee convergence. The proof of this theorem may be found in the reference [ORT70].

3.2.3. Waveform Relaxation

The relaxation schemes presented above can be also extended to functions spaces to solve systems of differential equations. This class of algorithms is called *Waveform Relaxation* (WR) [LEL82]. The relaxation variables in WR are elments of function spaces, i.e., they are waveforms in the closed interval [0,T], whereas for linear and nonlinear relaxation the variables are simply vectors in Euclidean n-space. To illustrate the WR algorithm, consider the circuit simulation problem in the form specified in Eq. (2.9). The WR method for solving this system of equations is given in Algorithm 3.5 below.

Algorithm 3.5 converts the problem of solving a coupled system of n first-order ODEs to the problem of solving n separate differential equations, each containing a single variable. The outer loop in the algorithm is the Gauss-Seidel iteration which requires that the latest values of the relaxation variables be used to solve each equation in the inner loop. Each equation in the inner loop is a single nonlinear differential equation, and this equation can be solved using any standard numerical integration method.

Algorithm 3.5 (WR Gauss-Seidel Algorithm for Solving Eq. (2.9))

$k \leftarrow 0$;
guess waveform $x^0(t)$; $t \in [0,T]$ such that $x^0(0) = x_0$;
repeat {
    $k \leftarrow k+1$;
    foreach ($i \in \{1,..,n\}$) {
        solve

$$\sum_{j=1}^{i} C_{ij}(x_1^k, \cdots, x_i^k, x_{i+1}^{k-1}, \cdots, x_n^{k-1}, u)\dot{x}_j^k +$$

$$\sum_{j=i+1}^{n} C_{ij}(x_1^k, \cdots, x_i^k, x_{i+1}^{k-1}, \cdots, x_n^{k-1}, u)\dot{x}_j^{k-1} +$$

$$f_i(x_1^k, \cdots, x_i^k, x_{i+1}^{k-1}, \cdots, x_n^{k-1}, u) = 0$$

for ($x_i^k(t)$; $t \in [0,T]$), with the initial condition
$x_i^k(0) = x_{i_0}$;

    }
} **until** ($\max_{1 \le i \le n} \max_{t \in [0,T]} | x_i^k(t) - x_i^{k-1}(t) | \le \epsilon$)

∎

The convergence of the Waveform Relaxation method is guaranteed under conditions which are similar to the linear and nonlinear cases, as stated in the following theorem [WHI85C]:

Theorem 3.3: If $C(x(t),u(t)) \in \mathbb{R}^{n \times n}$ of Eq. (2.9) is strictly diagonally dominant uniformly over all $x(t) \in \mathbb{R}^n$ and $u(t) \in \mathbb{R}^r$ and Lipschitz continuous with respect to $x(t)$ for all $u(t)$, then the sequence of waveforms $\{x^k\}$ generated by the Gauss-Seidel or Gauss-Jacobi WR algorithm will converge uniformly to the solution of Eq. (2.9) in any bounded interval $[0,T]$, for any initial guess $x^0(t)$. ∎

While this theorem guarantees convergence of the WR algorithm, it does not imply anything about the speed of convergence. Although the method usually converges in a few iterations, it has been observed that in test cases with tight feedback loops, the number of iterations required to converge is proportional to the simulation interval [WHI83]. To improve convergence, the simulation interval $[0,T]$ is usually divided into smaller intervals, $[0,T_1]$, $[T_1,T_2]$, ... , $[T_{n-1},T_n]$, called *windows*. Initially, the WR algorithm is applied only in the first window, $[0,T_1]$, until the waveforms converge. Then a second window, $[T_1,T_2]$, is selected and WR is applied within this interval until the waveforms converge. This continues until the entire simulation interval is covered. Note that the WR method converges more rapidly as the window size is made smaller. One advantage of WR is that the time-steps for each of the variables can be chosen independently of one another, but this advantage is compromised if the windows are too small. Therefore, the window size is an important factor which determines the performance of programs which use the WR method.

3.2.4. Partitioning for Relaxation Methods

Relaxation methods are most effective when applied to a system of equations which are "loosely-coupled," that is, where the variables do not depend too strongly on one another. For this type of system, relaxation methods usually converge quite rapidly. The speed of convergence in the linear case is controlled by the spectral radius of the iteration matrix given by $\rho(B^{-1}C)$ (using the notation of Theorem 3.1); this is usually close to zero for loosely-coupled systems. However, for an arbitrary problem, there is no guarantee that the spectral radius will be small. In fact, in "tightly-coupled" systems, the spectral radius may be very close to 1 which implies slow convergence. This degrades the performance of the relaxation-based methods compared to those for the direct methods.

The precise meaning of loosely-coupled and tightly-coupled can be described using a simple 2x2 matrix problem:

$$\begin{bmatrix} a_{11} & a_{12} \\ a_{21} & a_{22} \end{bmatrix} \begin{bmatrix} x_1 \\ x_2 \end{bmatrix} = \begin{bmatrix} b_1 \\ b_2 \end{bmatrix}$$

Assume that the equations have been ordered such that x_1 is solved before x_2. Then, a_{21} can be considered as a feed-forward term and a_{12} can be considered as a feedback term. The spectral radius of the iteration matrix for the GS method (see Theorem 3.1) is given by

$$\rho(B^{-1}C) = \left| \frac{a_{12}a_{21}}{a_{11}a_{22}} \right|$$

and to guarantee convergence, this value must be strictly less than 1. If both a_{12} and a_{21} are non-zero, the variables x_1 and x_2 are considered to be coupled. If both a_{12} and a_{21} are large, relative to a_{11} and a_{22}, then x_1 and x_2 are called *tightly-coupled* variables. If both a_{12} and a_{21} are small, then x_1 and x_2 are called *loosely-coupled* variables. Note that if either a_{21} or a_{12} is zero, then equation ordering has a significant impact on the number of iterations. In fact, if $a_{21}=0$, then x_2 should be solved before x_1 so that the solution can be obtained in one iteration. A similar argument applies if a_{21} is very small compared to a_{12}. Therefore, the main objective in reordering is to make the A matrix as lower triangular as possible.

When solving large systems, the definitions given above can be used to partition the system into groups of tightly-coupled variables. Rather than using relaxation methods to solve the tightly-coupled variables within each "block," it is better to solve them using direct methods. The relaxation method can be applied between the blocks, which are loosely-coupled relative to the variables within a block. This gives rise to block relaxation methods [VAR62], which can be viewed as a combination of the direct methods and relaxation methods. As an example, consider the 3x3 matrix problem:

$$\begin{bmatrix} a_{11} & a_{12} & 0 \\ 0 & a_{22} & a_{23} \\ 0 & a_{32} & a_{33} \end{bmatrix} \begin{bmatrix} x_1 \\ x_2 \\ x_3 \end{bmatrix} = \begin{bmatrix} b_1 \\ b_2 \\ b_3 \end{bmatrix}$$

If x_2 and x_3 are tightly-coupled, then many relaxation iterations may be required to solve this problem. However, by grouping x_2 and x_3 into the same block and reordering the variables for the Gauss-Seidel method, the following equation is obtained:

$$\begin{bmatrix} a_{22} & a_{23} & 0 \\ a_{32} & a_{33} & 0 \\ a_{12} & 0 & a_{11} \end{bmatrix} \begin{bmatrix} x_2 \\ x_3 \\ x_1 \end{bmatrix} = \begin{bmatrix} b_2 \\ b_3 \\ b_1 \end{bmatrix}$$

If x_2 and x_3 are solved using direct methods, then this problem can be solved using a single relaxation iteration. This example shows that proper ordering and partitioning are extremely important in the relaxation-based methods.

CHAPTER 4

ITERATED TIMING ANALYSIS

In the previous two chapters, the circuit simulation problem was identified and efficient techniques to solve the problem were described. In this chapter, a detailed description of event-driven electrical simulation based on nonlinear relaxation methods is provided. The chapter begins with the equation flow for nonlinear relaxation when applied to the circuit simulation problem. Then the timing analysis and iterated timing analysis (ITA) algorithms are described. An algorithm for event-driven electrical simulation with a global variable time step approach is given in the next section. Finally, the issues relating to latency detection and event scheduling in ITA are discussed.

4.1. EQUATION FLOW FOR NONLINEAR RELAXATION

The starting point for the description is the system of nonlinear differential equations describing the circuit behavior using the charge-based formulation:

$$\dot{q}(v(t)) = f(v(t), u(t)), \quad v(0) = V, \quad t \in [0, T] \tag{4.1}$$

where q is the charge associated with the capacitors connected to each node, f is the sum of the currents charging the capacitances at each node, u is the set of input voltages and v is the set of unknown node voltages. Using trapezoidal integration [CHU75] to discretize the system in Eq. (4.1), the following system of nonlinear difference equations is obtained:

$$q_{n+1} = q_n + \frac{h_n}{2}(f_{n+1} + f_n) \tag{4.2}$$

where the subscripts n and $n+1$ refer to time points t_n and $t_{n+1} = t_n + h_n$, respectively, and h_n is the integration step size. This equation can be formulated as a nonlinear problem, as follows:

$$F(v) = \frac{2}{h_n}(q_{n+1} - q_n) - (f_{n+1} + f_n) = 0 \qquad (4.3)$$

Instead of solving this system of equations using standard techniques [NAG75], the strategy in this section is to use nonlinear relaxation. That is, use the Newton method to solve each equation in the system separately and a relaxation method to guarantee that the solutions are mutually consistent. The expression for the ith equation in Eq. (4.3) solved using the Newton method is

$$J_{F_i}(v^k)(v_i^{k+1} - v_i^k) = - F_i(v^k) \qquad (4.4)$$

where the index k is the iteration counter for the Newton method and $J_{F_i}(v)$ is the ith diagonal term of the Jacobian matrix of $F(v)$ given by

$$J_{F_i} = \frac{2}{h}\frac{\partial q_i(v^k)}{\partial v_i} - \frac{\partial f_i(v^k)}{\partial v_i} \qquad (4.5)$$

Usually a number of iterations are required to obtain the correct solution. However, in this case, since a converged relaxation method is used to guarantee a consistent solution to the system of equations, the Newton iteration for each equation need not be carried to convergence. In fact, from an efficiency standpoint, *only one iteration* should be used to approximate the solution of each equation before moving to the next equation, as described earlier in Chapter 3. The resulting one-step Gauss-Seidel-Newton relaxation algorithm is specified precisely in the following, using the definition:

$$v^{k,i} = [v_1^{k+1}, v_2^{k+1}, \cdots, v_{i-1}^{k+1}, v_i^k, v_{i+1}^k, \cdots, v_n^k]^T$$

where the superscript T denotes the transpose of a vector. This definition is based on the Gauss-Seidel method which uses the $k+1^{st}$

values of all other components, whenever possible, in computing the $k+1^{st}$ value of v_i. Here n is the number of equations in the system.

Algorithm 4.1: (Gauss-Seidel-Newton Relaxation Method)
> **repeat** {
> > **foreach** ($i \in \{1, \cdots ,n\}$) {
> > > solve $J_{F_i}(v^{k,i})(v_i^{k+1} - v_i^k) = - F_i(v^{k,i})$ for v_i^{k+1}
> > > where $F_i(v)$ is specified in Eq. (4.3) and
> > > $J_{F_i}(v)$ is specified in Eq. (4.5) ;
> >
> > }
> } **until** ($|| v_i^{k+1} - v_i^k || < \varepsilon_1$, $|| F_i || < \varepsilon_2$, i=1,..,n)

∎

4.2. TIMING ANALYSIS ALGORITHMS

The first published program to use techniques based on nonlinear relaxation for circuit simulation was the MOTIS program [CHA75]. It used backward-Euler integration, a Gauss-Jacobi-Newton relaxation algorithm, and node-by-node decomposition (that is, it solved for one node voltage at a time). In MOTIS, a simple modification was made to the relaxation scheme based on the conjecture that there exists a small enough time-step, h_{min}, such that the method obtains the correct solution in exactly one iteration. At each time point, t_{n+1}, the program computed new values of all node voltages using only one iteration of the Gauss-Jacobi-Newton method and accepted the results as the correct solutions at t_{n+1}. It was believed that iterating the outer relaxation loop to convergence would be both expensive and unnecessary for most MOS logic circuits. However, the resulting accuracy of this approach relied heavily on three things:

(1) The user's ability to select an appropriate time-step based

on knowledge of the circuit characteristics

(2) The fact that the global error reduces to zero when a node voltage reaches the supply voltage or ground

(3) Only a limited number of well-characterized circuit topologies (CMOS polycells) were used to build a design.

The initial speed improvements obtained using this approach were extremely encouraging, partially due to the simplified numerical techniques and partially due to the use of table lookup models for the MOS devices. The combined techniques were shown to be over two orders of magnitude faster than standard techniques when applied to large digital MOS circuits [CHA75]. Since the method was intended to provide first-order timing information of MOS logic circuits, it was called "Timing Analysis" or "Timing Simulation."

Although timing analysis provided an electrical simulation capability with execution speeds comparable to logic simulation, it had a number of problems. For example, the choice of a proper time-step to guarantee accurate solutions was very difficult to determine in general. In addition, the method had severe accuracy problems for circuits containing elements such as large floating capacitors[1], small floating resistors and transfer gates. The MOTIS program avoided this problem for floating capacitors by not allowing them in the circuit description and solved collections of transfer gates using direct methods.

A number of improvements to the basic technique was suggested to overcome the inherent accuracy limitations of the method. In particular, the MOTIS-C program [FAN77] employed trapezoidal integration and one iteration of the Gauss-Seidel-Newton relaxation algorithm. Since

---

[1] A "floating" element is a two-terminal device whose terminals are not connected to either ground or to a power supply.

timing analysis algorithms based on the Gauss-Seidel principle use updated information at t_{n+1} whenever possible, the accuracy is generally better than one based on the Gauss-Jacobi method. The simulation time-step was selected automatically in the program by doing a simple analysis of the time constants associated with each node and by using some fraction of the smallest time constant as the step size. However, MOTIS-C still suffered from problems similar to those for MOTIS.

A modified timing analysis algorithm was implemented in SPLICE1.3 [NEW78] as part of a mixed-mode simulation capability. Although backward-Euler integration was used in this program, a number of other noteworthy enhancements were made to the underlying timing analysis algorithm. The first enhancement was based on two observations:

(1) Most of the node voltages in a large digital circuit remain stationary at a given time point (the latency property). Computing the solution for these nodes is unnecessary.

(2) The order in which the nodes are solved has a strong influence on the accuracy of the solution for timing analysis algorithms based on the Gauss-Seidel principle.

These observations suggested that a good strategy would be to identify the "active" nodes at each time point and process these nodes in an order based on the direction of signal flow. In SPLICE1.3, a single mechanism was used to perform both tasks: an *event-driven, selective-trace* algorithm normally associated with logic simulation [SZY75]. This mechanism is described in the following paragraphs.

The SPLICE1 program treats a circuit as a signal-flow graph and constructs a corresponding directed graph for the circuit given by $G=G(X,E)$, where X is the set of vertices and E is the set of directed edges of the graph. Two tables, the *fanin* and *fanout* tables, are constructed at each vertex based on the following definitions:

Definition 4.1: (Fanin and Fanout nodes)

A node x_k is called a fanin node of x_l, and is specified as $x_k \in$ **Fanin**(x_l), if x_k directly affects x_l. A node x_j is called a fanout node of x_i, and is specified as $x_j \in$ **Fanout**(x_i), if x_j is directly affected by x_i. ∎

Whenever the value of an input node or any internal node changes, it is possible to *schedule* all of its fanouts to be processed. In this way the effect of a change at the input to a circuit may be *traced* as it propagates to other circuit nodes via the fanout tables. Since the only nodes that are processed are those which are affected directly by the change, this technique is *selective* and hence its name: *selective trace*. If such a selective trace algorithm is used with the fanout tables, the order in which the nodes are updated becomes a function of the signals flowing in the network and is therefore a dynamic ordering.

To make the processing efficient, and for consistency with the logic simulator in the SPLICE1 program, the total simulation period, T_{stop}, is divided into uniform steps, referred to as the *Minimum Resolvable Time* (**mrt**). A time queue is constructed and the time slots in this queue define distinct points in time separated by one **mrt**. Hence, events are scheduled at integer multiples of **mrt** in the queue. The simple event scheduling algorithm used in SPLICE1 for timing analysis is given below. The routine *NextEventTime*(t) examines successive time slots in the time queue starting at time t and returns the next time point where one or more events have been scheduled. The external input nodes to a circuit are denoted as e_k.

As seen in the algorithm below, three separate event scheduling mechanisms exist:

(1) External inputs generate events whenever they make transitions from one value to another,

(2) Internal nodes can schedule themselves to be processed, and

(3) Internal nodes can schedule their fanout nodes to be processed.

Note that if x_i is not active, then neither x_i nor its fanouts are scheduled. However, since nodes may schedule themselves, the fanouts of x_i may still be active even though x_i is not. The importance of this fact and other issues associated with electrical event scheduling will be presented in Section 4.6. Also, the precise meaning of "active" is elaborated further in Section 4.6.

Algorithm 4.2: (Event Scheduling Algorithm in SPLICE1)

```
tn← 0;
while ( tn≤Tstop) {
    tn← NextEventTime( tn );
    foreach ( input k at tn ) {
        if ( ek is "active" )
            forall ( xj ∈ Fanout(ek) ) schedule( xj, tn );
    }
    foreach ( event i at tn ) {
        process node xi by computing xi(tn);
        if ( xi is "active" ) {
            schedule( xi, tn+h );
            forall ( xj ∈ Fanout(xi) ) schedule( xj, tn );
        }
    }
}
```
∎

The use of event-driven, selective-trace techniques give greatly improved accuracy of SPLICE1.3 compared to those for the MOTIS and MOTIS-C programs. In addition, a further improvement was realized using a variable time-step control, as follows. Initially, every node is solved using a common step size given by the **mrt**. If the change in either the voltage at a node or the current through any device connected to the node is large, its solution is recomputed in the **mrt** interval using smaller steps and a single iteration at each time point. Each of the smaller steps may be further refined to insure that the changes in voltage and current are within acceptable limits. Therefore, the local time-steps for each node are based on limiting change of the node voltage and its associated currents over each step[2]. While the run time was noticeably higher, this variable time-step control was extremely effective in improving the accuracy of the results.

Other enhancements were developed in SPLICE1.3 to handle tightly-coupled circuits. SPLICE1.3 used the Implicit-Implicit-Explicit (IIE) method [NEW80] to handle floating capacitors. To accommodate large blocks of tightly-coupled circuit elements, the program allowed the user to define "circuit" blocks. These blocks would be solved using standard direct matrix techniques. However, instead of using a single iteration, the Newton iteration in the inner loop was carried to convergence since the elements inside the circuit block were considered to be "highly" nonlinear. However, the outer relaxation iteration was only performed once.

While the results from programs using timing analysis were within acceptable accuracy limits for a certain class of problems, a rigorous mathematical analysis indicated that these methods have inherent stability and accuracy problems [DEM81B]. This severely limited the application of the technique. Another problem, cited earlier, was that timing analysis programs relied on the user's knowledge of the underlying

---

[2] Note that a variable time-step control based on local truncation error is not easy to define here since the relaxation loop is not carried to convergence. The local error (i.e., the error over one step) is due to the integration method and the fact that the iteration is not carried to convergence.

algorithms and improper usage could produce the wrong answer. Circuit designers have been known to lose confidence in a simulator if it occasionally produces the wrong answer, whatever the reason. Therefore, this approach has not been widely accepted, although it is heavily used where the approach has been thoroughly developed, is well-understood, and is applied to a restricted class of circuit topologies.

4.3. SPLICE1.7 - FIXED TIME-STEP ITA

The reluctance to close the outer relaxation loop in timing analysis was primarily due to its perceived high cost. However, the event-driven techniques significantly reduced the cost of timing analysis for large problems since only a small fraction of the nodes is processed at each time point. A number of other improved timing analysis algorithms were proposed [DEM83] but they used at least two iterations or required the use of expensive function evaluations, which increased greatly the cost of the simulation. As described earlier, the variable step approach in SPLICE1.3 improved the accuracy somewhat at the expense of additional iterations. The additional cost was thought to be worthwhile due to the improved reliability.

The next step, naturally, is to close the relaxation loop and examine the true cost of iterating to convergence, given that event-driven selective trace is employed to improve efficiency. This was done in the SPLICE1.6 program, which later evolved to be SPLICE1.7, and the technique was named Iterated Timing Analysis or ITA [SAL83]. The prototype version of ITA used backward-Euler integration, node-by-node decomposition and a fixed time-step based on the **mrt**. The fixed time-step algorithm was kept for consistency with the existing scheduler and logic simulation portions of SPLICE1. The ITA algorithm in SPLICE1.7 is a simple extension of Algorithm 4.3 as shown below.

Algorithm 4.3: (Fixed Time-Step ITA)

```
t_n ← 0;
while ( t_n ≤ T_stop ) {
    t_n ← NextEventTime( t_n );
    foreach ( input k at t_n )
        if ( e_k is active )
            forall ( v_j ∈ Fanout(e_i) ) schedule( v_j, t_n );
    repeat {
        foreach ( event i at t_n ) {
            solve J_{F_i}(v^{k,i})(v_i^{k+1} - v_i^k) = - F_i(v^{k,i}) for v_i^{k+1}
                where F_i(v) is specified in Eq. (4.3) and
                    J_{F_i}(v) is specified in Eq. (4.5) ;
            if ( |v_i^{k+1} - v_i^k| < ε_1,  |F_i| < ε_2 ) {  /* converged? */
                if ( v_i did not converge on last iteration ) {
                    if ( v_i is active ) {
                        /* this is the selective-trace portion */
                        schedule( v_i, t_{n+1} );
                        forall ( v_j ∈ Fanout(v_i) )
                            schedule( v_j, t_n );
                    }
                    else {/* do nothing (latency) */}
                }
                else {/* do nothing (break feedback loops) */}
            }
            else {  /* node has not converged */
                schedule( v_i, t_n );
                forall ( v_j ∈ Fanout(v_i) ) schedule( v_j, t_n );
            }
        }
    } until ( Q is empty at t_n )
}
```

The following definition is used above:

$$v^{k,i} = [v_1^{k+1}, v_2^{k+1}, \cdots, v_{i-1}^{k+1}, v_i^k, v_{i+1}^k, \cdots, v_n^k]^T$$

The algorithm above has two features not present in the SPLICE1.3 algorithm:

- If a node voltage does not converge, the node is rescheduled at the current time point t_n along with its fanout nodes.

- All nodes are processed until their voltages converge. When a node converges at t_n, it schedules itself at t_{n+1} and schedules its fanouts at t_n, if active. However, if it is scheduled again at t_n, by one of its fanins, and converges again, it does not schedule any additional events. This approach breaks feedback loops, since two nodes which are fanouts of each other would schedule each other indefinitely at t_n if this approach was not used.

The speed improvement obtained by the SPLICE1.7 program compared to that for the SPICE2 program was in the range of 5 to 50 times faster for a number of MOS digital circuits containing up to 1200 transistors [SAL83]. However, the ITA approach required approximately twice as much CPU-time to simulate a circuit compared to SPLICE1.3 which used timing simulation [SAL84]. Again, the improvements in reliability and numerical robustness far outweighed the cost of the increase in run-time.

While the converged relaxation scheme is provably better than the non-iterated approach, it is not without problems. One problem is the speed of convergence. For example, SPLICE1.7 was able to simulate accurately an NMOS operational amplifier but it required more than two times the CPU-time used by SPICE2 [NEW83]. The circuit is a tightly-coupled analog circuit with large forward gain and capacitive feedback and, in this application, the node-by-node decomposition strategy used in SPLICE1.7 is inappropriate. For this same reason, convergence is also very slow in the presence of large floating capacitors and small drain and source resistors, usually found in detailed MOS transistor models. Another problem is due to nonconvergence. Since a fixed time-step is used, the program simply stopped when it was unable to converge to a solution within a specified number of relaxation-Newton iterations. Obviously, a variable step algorithm would resolve this problem and would also allow the solutions to be computed accurately based on a local truncation error criterion. These and other problems were solved in the SPLICE2 and iSPLICE3.1 programs.

4.4. iSPLICE3.1 - GLOBAL-VARIABLE TIME-STEP ITA

A new robust version of ITA has been implemented in the iSPLICE3 [SAL89A] program. It differs from SPLICE1.7 in two respects:

- it uses partitioning to improve the speed of convergence for tightly-coupled circuits

- it achieves better accuracy by using an LTE-based time-step control.

The iSPLICE3 program also provides detailed MOS level 1 and MOS level 3 transistor models including a charge-conserving capacitance model.

4.4.1. Circuit Partitioning

The node-based ITA approach used in SPLICE1.7 is not appropriate for circuits with tight coupling between two or more nodes, since the convergence can be very slow in this situation. One reason for this problem is that, in computing the new value for a particular node, the relaxation process effectively replaces the fanin nodes with ideal voltage sources of constant value. Therefore, the true Norton equivalent contributions from the fanin nodes are not used in the computation of a new value for the node. SPLICE2 used an improved representation of the neighboring nodes based on a current and conductance model, rather than constant voltage sources, and this approach was called the *coupling method* [KLE84]. This fanin node model is only approximate since the exact Norton equivalent circuit for each node is expensive to calculate for large circuits. While this approach improved the convergence speed on some examples, the technique was heuristic in nature and did not solve the general problem of coupling between more than two nodes in feedback loops.

As was realized in early mixed-level simulators such as SPLICE1, tightly-coupled subcircuits are better solved using direct methods [NEW78]. However, it is difficult for users to identify tightly-coupled blocks manually, especially when the degree of coupling is a function of time and hence may change over the simulation interval. A more effective approach to the coupling problem is to identify strongly-coupled components in the circuit automatically and to group them together to form subcircuits - a process referred to as *circuit partitioning*. Since the variables associated with the subcircuits are assumed to be tightly-coupled, the subcircuits can each be solved using direct matrix techniques, and the relaxation method can be applied between subcircuits. This technique has been used in conjunction with the Waveform Relaxation algorithm [LEL82, CAR84, WHI85A, MAR85, DUM86] with great success. The same approach can be used with nonlinear relaxation to improve convergence as described in Chapter 3. The static partitioning approach of the RELAX2 program [WHI85C] has been adopted in the iSPLICE3 program and it is described briefly in the following.

The main goal of partitioning is to speedup the convergence process of relaxation methods. Recall from Chapter 3 that the speed of convergence is controlled by the contraction factor, γ_∞, in the following way:

$$|| x^{k+1} - x^k || \leq \gamma_\infty || x^k - x^{k-1} ||$$

For a linear problem, this iteration factor can be computed quite easily. For example, if the linear problem $\mathbf{Ax=b}$ is solved using the Gauss-Seidel algorithm, γ_∞ is equal to the largest eigenvalue of the iteration matrix $[(\mathbf{L+D})^{-1}(-\mathbf{U})]$, where $\mathbf{A=L+D+U}$. Therefore, a two-node linear circuit, such as the one in Fig. 4.1, has an iteration factor (for the conductance portion only) given by

$$\gamma_\infty = \frac{g_{12}}{(g_2+g_{12})} \frac{g_{12}}{(g_1+g_{12})}$$

A similar expression exists for the capacitance portion of the circuit. Note that if the two nodes are part of a larger circuit, the values of g_1 and g_2 are the Norton equivalent conductances seen from each node looking back into the rest of the circuit.

The partitioning algorithm makes use of the iteration factor to decide whether or not two nodes should be placed in the same subcircuit. If the factor is close to one and the nodes are solved independently, the convergence would be very slow. Therefore, the nodes should be placed in the same subcircuit. However, if the factor is close to zero, they may be placed in different subcircuits without adversely affecting the convergence speed. A threshold parameter, α, is used to decide whether or not the nodes should be solved together or separately.

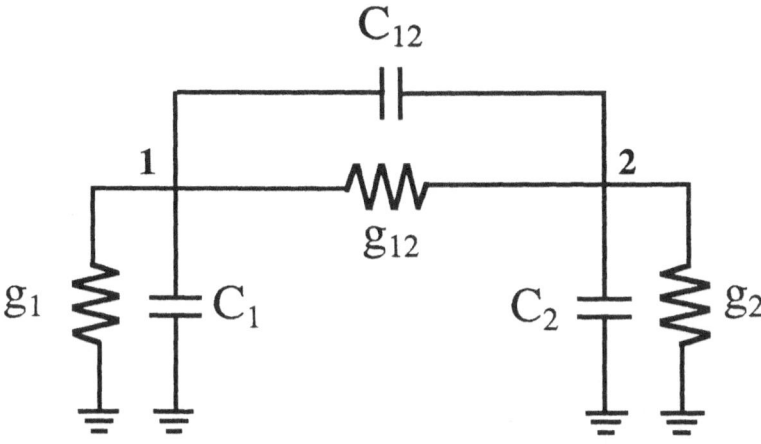

Figure 4.1: Linear Circuit Considered for Partitioning Purposes

A number of approximations are made in computing the iteration factors when partitioning MOS circuits. As MOS circuits are nonlinear, each nonlinear device must be replaced by a linear equivalent device. Since a static partitioning strategy is used, worst-case conductance and capacitance values are used when replacing each nonlinear device with a linear one. However, the exact Norton equivalent model seen by each node cannot be computed efficiently because it involves tracing paths from each node to all other nodes in the circuit. For efficiency, the depth of the conductance and capacitance computing processes is truncated whenever the gate of an MOS transistor is encountered since the conductance of an MOS transistor is zero in the worst case. When these heuristics are applied, the following partitioning algorithm is obtained:

Algorithm 4.4 (Conductance Partitioning)

$g_{12} \leftarrow 0$; $g_1 \leftarrow 0$; $g_2 \leftarrow 0$;
foreach (conductive element between nodes 1 and 2) {
   $g_{12} \leftarrow g_{12} +$ maximum element conductance over all v;
   Remove the element from the circuit;
}
$g_1 \leftarrow$ sum of the minimum Norton equivalent
  conductance of each element at node 1
$g_2 \leftarrow$ sum of the minimum Norton equivalent
  conductance of each element at node 2
if ($\dfrac{g_{12}}{(g_2+g_{12})} \dfrac{g_{12}}{(g_1+g_{12})} > \alpha$) {
   Place the two terminal nodes in same subcircuit;
}

■

A similar algorithm is used for partitioning based on capacitances. Using this approach, the run times were reduced significantly compared to those for the node-based approach on all examples simulated. However, the partitioning strategy described here has a number of problems. The main problem with this approach is that it may produce unnecessarily large subcircuits since worst-case values are used in the partitioning process. The advantages of the relaxation method are lost if the subcircuits are too large. Since static partitioning is used (that is, the subcircuits are defined before the simulation begins), the latency exploitation is no longer performed at the node level but rather at the subcircuit level. All nodes in a subcircuit must be latent before the subcircuit is declared latent. While this provides a somewhat stronger condition for latency, it reduces the efficiency of the latency exploitation. Ideally, one would prefer to use small-signal conductance and capacitance values to perform the initial partitioning, and then adjust the subcircuits as these values change during the simulation. This is referred to as *dynamic partitioning* and has already been successfully applied to the simulation of bipolar circuits using Waveform Relaxation [MAR85].

Another problem with the partitioning approach given in Algorithm 4.4 is that it is too local a criterion. For example, if two nodes are extremely tightly-coupled, relative to their coupling to neighboring nodes, they will be placed in the same subcircuit while the neighboring nodes may be incorrectly placed in different subcircuits. If the neighboring nodes are actually coupled to either of the two external nodes, the convergence will still be slow [WHI85C]. One practical problem in partitioning is that it is a time-consuming task. Care must be taken in the definition of the data structures and partitioning algorithms so that the partitioning phase does not dominate the total run time for large circuits. This is more of a concern in dynamic partitioning [MAR85] where the partitioning operation may be performed frequently during the simulation.

4.4.2. Global-Variable Time-Step Control

iSPLICE3.1 uses a global-variable time-step algorithm in which the components in the system are integrated using a single common time-step. This integration time-step is selected based on the fastest changing variable in the system, the same strategy used in direct methods. However, only the active subcircuits are processed at each time point, and these subcircuits are identified using the selective-trace algorithm. The main steps in the global time-step ITA algorithm are given below following a brief description of the notation to be used.

Notation for Algorithm 4.5: (see Fig. 4.2)
Assume that a given circuit is partitioned into n subcircuits $S_1, S_2, \cdots, S_i, \cdots, S_n$. The ith subcircuit, S_i, has n_i internal variables

and n_e external inputs. The internal variables given by int(S_i) = { x_1 , x_2, ... , x_{n_i} } are those variables computed whenever subcircuit S_i is processed. They are defined in vector form as $v_i = [x_1, x_2, \cdots, x_{n_i}]^T$. The external inputs of a subcircuit are other nodes which affect the internal nodes of the subcircuit. They are specified as **Fanin**(S_i) = { $e_1, e_2, \cdots, e_{n_e}$}. The fanouts of a subcircuit are associated with the internal nodes of the subcircuits. Hence, the set of subcircuits affected by an internal node, x_j, are specified as **Fanout**(x_j) = { $S_1, S_2, \cdots, S_k$}. The following definition is also used:

$$v^{k,i} = [v_1^{k+1}, v_2^{k+1}, \cdots, v_{i-1}^{k+1}, v_i^k, v_{i+1}^k, \cdots, v_n^k]^T.$$

Algorithm 4.5: Global-Variable-Time-Step ITA

```
partition();
tₙ← 0;  hₘᵢₙ← hₛₜₐᵣₜ;
while ( t ≤ Tₛₜₒₚ) {
    stepRejection = FALSE;
    hₙₑₓₜ← hₘᵢₙ; tₙ← tₙ + hₙₑₓₜ; hₘᵢₙ← hₘₐₓ;
    foreach ( input iₖ at tₙ )
        if ( eₖ is active )
            forall ( Sⱼ ∈ Fanout(eₖ))   schedule( Sⱼ, tₙ );
    repeat {
        foreach ( event i at tₙ ) {
            solve J_F_i(vᵏ⁼ⁱ)(vᵢᵏ⁺¹− vᵢᵏ) = − Fᵢ(vᵏ⁼ⁱ) for vᵢᵏ⁺¹
            corresponding to subcircuit Sᵢ;
            if ( || vᵢᵏ⁺¹− vᵢᵏ || <ε₁,  || Fᵢ || <ε₂ ) {  /*converged? */
                if ( vᵢ did not converge on last iteration ) {
                    foreach ( xᵢ ∈ int(Sᵢ) ){
                        if ( xᵢ is active ) {
                            if ( CheckAccuracy( xᵢ ) = TRUE ) {
                                hᵢ← pickStep( xᵢ );
```

```
                              h_min← min( h_min, h_i );
                              schedule( x_i, t_{n+1} );
                              forall (S_j ∈ Fanout(x_i))
                                  schedule( S_j, t_n );
                          }
                          else {  /* reject solution */
                              t_n← t_n− h_min;
                              h_min← h_min/2;
                              stepRejected = TRUE;
                          }
                      }
                  }
              }
          }
          else {  /* subcircuit has not converged yet */
              if ( itercnt > maxitercnt ) {
                  t_n← t_n− h_min; h_min← h_min/2;
                  stepRejected = TRUE;
              }
              else {
                  schedule( S_i, t_n );
                  foreach ( x_i ∈ int(S_i) ) {
                      if ( x_i is active )
                          forall ( S_j ∈ Fanout(x_i) )
                              schedule( S_j, t_n );
                  }
              }
          }
      }
  } until (( Q is empty at t_n ) OR (stepRejection) )
}
```

■

In the algorithm above, the *CheckAccuracy*(x) routine uses a local truncation error criterion to determine if the computed solution for x is

accurate and, if so, returns "TRUE." The *PickStep(*x*)* routine uses an LTE estimate to pick the next recommended step size for **x**.

The main differences between this algorithm and the one used in SPLICE1.7 are due to the actions taken when the subcircuit variables converge at a time point and when they do not converge in a specified number of relaxation-Newton iterations. When the active subcircuits converge at a time point, t_n, the local truncation errors for their internal variables are estimated [BRA72] and the new global time-step, h_{next}, is set to the smallest recommended step in the system, h_{min}. If the accuracy in the solution computed at t_n is unacceptable, the solution is rejected and the integration is retried with the smaller time-step. Similarly, if the iterations do not converge within a specified number of iterations, the time-step is rejected and a smaller step is used.

4.5. ELECTRICAL EVENTS AND EVENT SCHEDULING

4.5.1. Latency Detection

The most critical aspect in ITA, in terms of accuracy, is the detection of the latency condition. For example, if component **x** is identified as being latent prematurely, any small errors in its value will be propagated to the other components producing errors in their solutions. If the component is thought to be latent but, in reality, it is changing very slowly, the results may be completely wrong. Then the overriding question is: how can one be sure that a variable has reached a steady-state value? The simplest approach is to test if the following condition is satisfied:

Latency Condition 1:

$$|x_{n+1} - x_n| < \varepsilon_x \qquad\qquad (4.6)$$

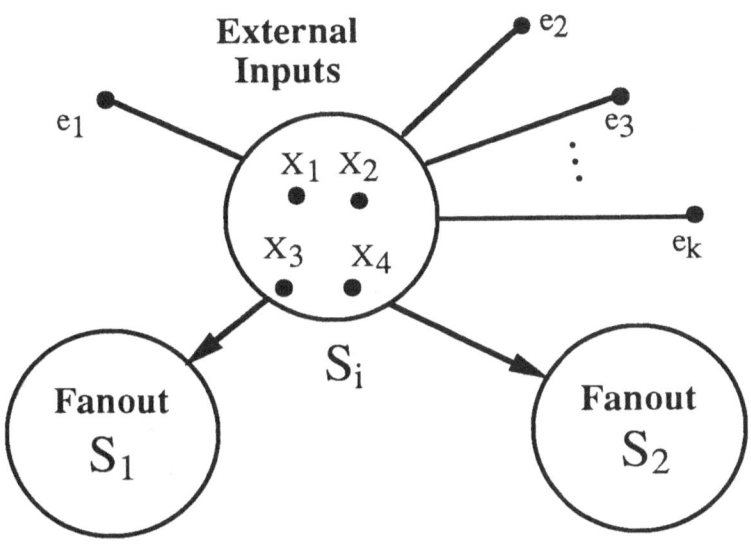

Figure 4.2: Notation Associated with Subcircuits

where $x_{n+1} = x(t_{n+1})$, $x_n = x(t_n)$ and ε_x is some small number. As illustrated in Fig. 4.3, the component is considered latent if the difference in the computed solution at two successive time points is less than some pre-specified amount, ε_x. For a fixed time-step ITA algorithm [SAL83], this is a reasonable check as long as ε_x is specified properly and one additional check is done, as described shortly. There are situations where Condition 1 may fail, as shown in Fig. 4.4, where the true solution rises and then falls before reaching a steady-state value. If the time points are chosen such that Condition 1 is satisfied, latency will be detected incorrectly. A more conservative version of Condition 1 requires that the inequality be satisfied for two time points that are not adjacent.

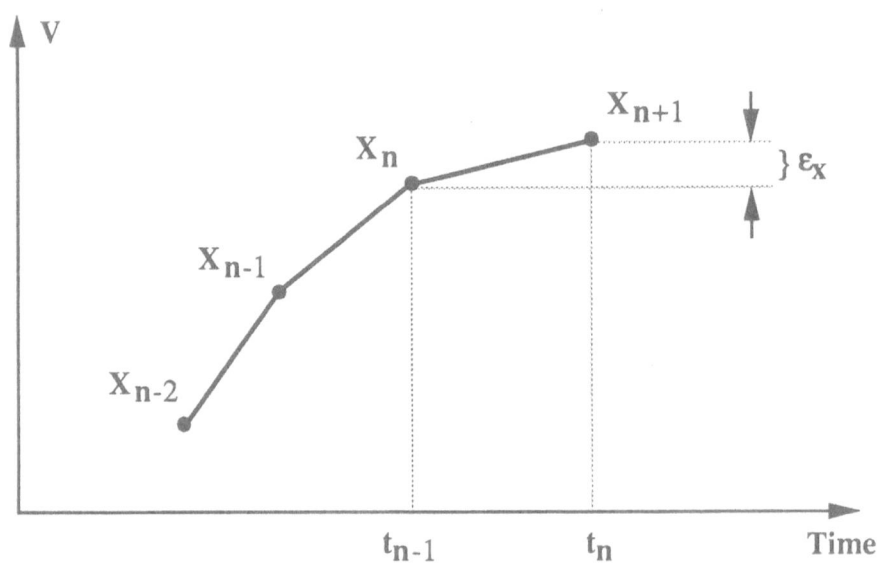

Figure 4.3: Simple Latency Detection

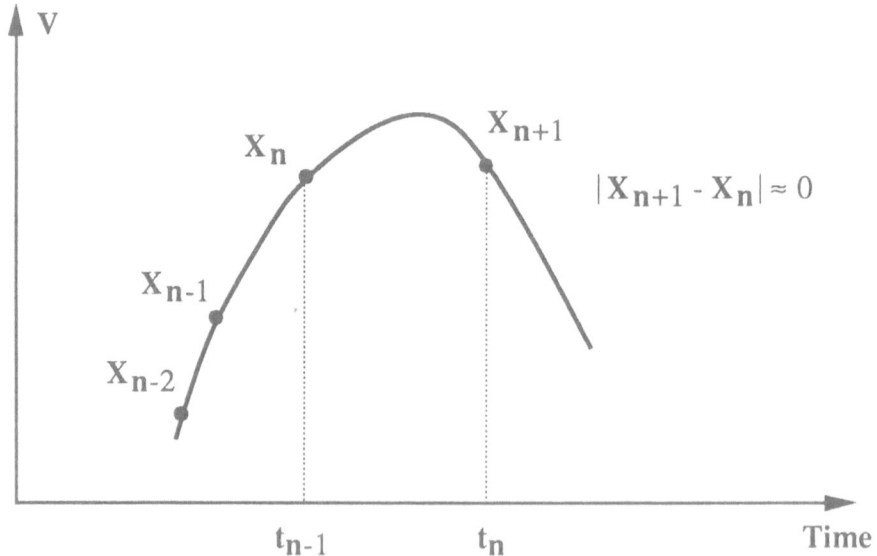

Figure 4.4: Potential Problem in Latency Detection

Latency Condition 1.1:

$$|x_{n+k} - x_n| < \varepsilon_x , \quad k>1 \tag{4.7}$$

While this conservative approach works well in practice, it is still not strong enough to handle the general case. For example, if a global variable time-step control is used, the step sizes may be very small due to some fast component resulting in small changes in x over a large number of time points (if x is a slower component). In this case, it would make more sense to use a rate-of-change criterion to detect latency rather than the absolute change in x. That is, use the check

Latency Condition 2:

$$\frac{|x_{n+1} - x_n|}{h_n} < \varepsilon_{\dot{x}} \tag{4.8}$$

As shown in Fig. 4.5, this requires that $\dot{x} \approx 0$ to satisfy the latency condition. This method also encounters problems with the example in Fig. 4.4 since $\dot{x} \approx 0$ as the signal switches direction. A more conservative way to do this type of latency check would be to use the strategy of Condition 1.1 and include a number of points from the past.

Latency Condition 2.1:

$$\frac{1}{k}\sum_{j=1}^{k} \frac{|x_{n+2-j} - x_{n+1-j}|}{h_{n+1-j}} < \varepsilon_{\dot{x}} , \quad k \geq 1 \tag{4.9}$$

This condition uses an average rate of change based on the previous k solutions to detect latency and this overcomes the problem given in Fig. 4.4. However, another problem arises if the true value of $\dot{x}$ is some small non-zero value that eventually changes the value of x significantly

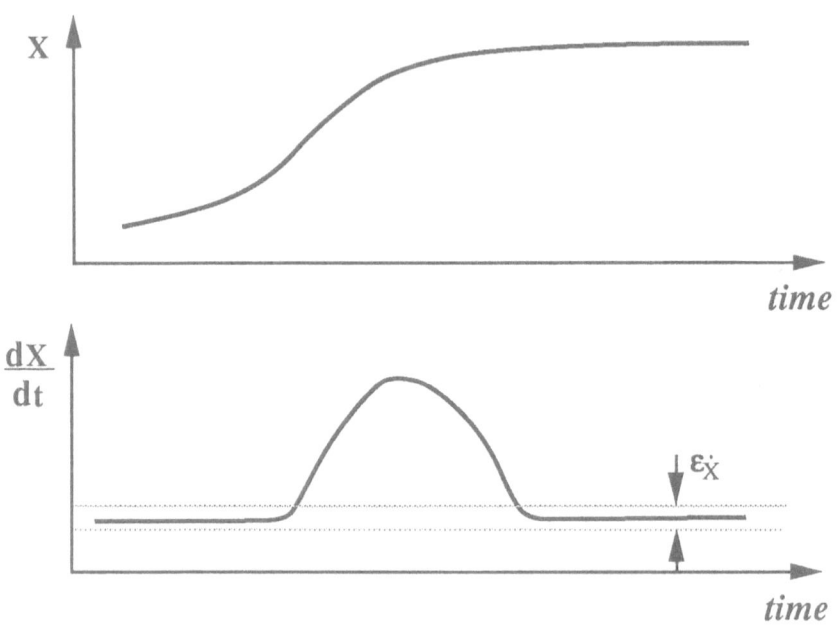

Figure 4.5: Variable Step Latency Criterion Based on Rate-of-Change

at some point in the future. To resolve this problem, a "wake-up" mechanism should be used with either Condition 1.1 or 2.1 when it is anticipated that component x has undergone a significant change in value. That is, the actual rate-of-change of x should be used to predict the wake-up time point, as follows:

Wake-up Condition 1:

$$h_{next}\frac{|x_{n+1} - x_n|}{h_n} > \varepsilon_x \qquad (4.10)$$

and $t_{wake-up} = t_{n+1} + h_{next}$. This wake-up condition can be used to compute h_{next} and the component should be re-activated and solved at $t_{wake-up}$. This process is illustrated in Fig. 4.6.

The latency and wake-up conditions specified above work well in practice and their use can be justified by considering latency exploitation as the use of a zeroth-order explicit integration method as described in reference [RAB79]. Explicit integration algorithms are obtained directly from a Taylor series expansion of the solution at the point t_n:

$$x_{n+1} = x_n + h_{n+1}\dot{x}_n + \frac{h_{n+1}^2}{2}\frac{d^2x_n}{dt^2} + \cdots \quad (4.11)$$

A zeroth-order method uses only the first term and produces the

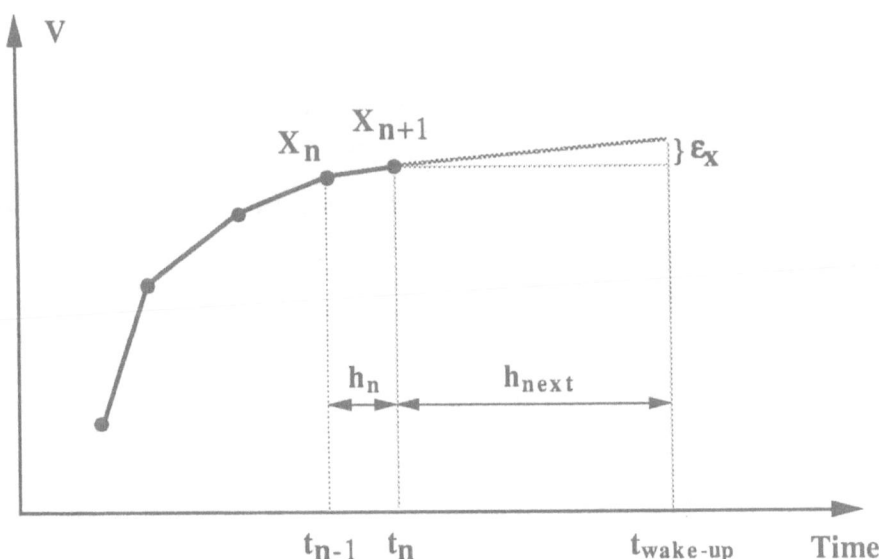

Figure 4.6: Wake-up Mechanism

following trivial integration method for which $x(t_{n+1})$ is simply updated with the value $x(t_n)$ at the previous time point:

$$x_{n+1} = x_n \tag{4.12}$$

This integration method has a local truncation error (LTE) given by

$$LTE = h_{n+1}\dot{x}(\xi) \qquad t_n \leq \xi \leq t_{n+1}$$

An estimate of the LTE can be obtained using a finite difference approximation for $\dot{x}$:

$$\dot{x}_n(\xi) \approx \frac{x_{n+1} - x_n}{h_n}$$

Therefore the LTE estimate is given by

$$LTE \approx h_{n+1}\frac{x_{n+1} - x_n}{h_n}$$

A check for latency can now be constructed from this analysis. The integration method specified in Eq. (4.12) can be used whenever the following condition is satisfied:

Latency Condition 3:

$$h_{n+1}\frac{|x_{n+1} - x_n|}{h_n} < E_{userLTE} \tag{4.13}$$

where $E_{userLTE}$ is the allowable local truncation error specified by the user.

For a fixed time-step algorithm, this latency check is equivalent to Condition 1 since $h_n = h_{n+1}$ for all n. Of course, the value for ε_x in Condition 1 must be derived the same way as $E_{userLTE}$ to be identical to

Condition 3. For a variable step algorithm, one could rewrite Condition 3 as

$$\frac{|x_{n+1} - x_n|}{h_n} < \frac{E_{userLTE}}{h_{new}}$$

By replacing h_{n+1} with a constant value of step size h_{max} such that $h_{max} \gg h_{n+1}$, one can provide a somewhat tighter constraint:

$$\frac{|x_{n+1} - x_n|}{h_{n+1}} < \frac{E_{userLTE}}{h_{max}}$$

Then latency condition 2 and 3 can be made identical by setting $\varepsilon_{\dot{x}} = E_{userLTE}/h_{max}$. Note that Condition 3 is an *a posteriori* criterion (i.e., it is used after selecting h_{n+1}) to detect latency. A similar criterion can be used in an *a priori* manner to decide when to activate the component. The idea is to use the LTE requirement to predict the time point when the zeroth-order integration method is no longer valid by checking when Latency Condition 3 is violated:

$$h_{new} \frac{|x_{n+1} - x_n|}{h_n} > E_{userLTE} \qquad (4.14)$$

where $h_{new} = t_{wake-up} - t_{n+1}$ and $t_{wake-up}$ is the time when the component should be activated. This wake-up time can be computed as follows:

$$t_{wake-up} = t_{n+1} + \frac{E_{userLTE}h_{n+1}}{x_{n+1} - x_n} \qquad (4.15)$$

and this is identical to Wake-up Condition 1. Therefore, the intuitive arguments which lead to Latency Conditions 1 and 2 and Wake-up Condition 1 are well-supported by the above analysis.

4.5.2. Events and Event Scheduling

The next issue is to define precisely the notion of electrical events for use in conjunction with the scheduling algorithm. The proper definition of this concept is important from the standpoint of efficiency and accuracy, as will be seen. In logic analysis, an event occurs when a node makes a transition from one state to another (different) state. The event causes the fanouts of the node to be scheduled in the time queue. As long as the node remains in the same state, no additional events are generated. Since logic states are discrete, logic events are easy to identify. In electrical analysis, there is a continuum of "allowed states" making it more difficult to distinguish a significant event from an insignificant one. However, the definition of logic events can be extended in a straightforward manner to electrical analysis. The resulting definition of an electrical event is connected with the notion of "active" and "latent" components. <u>Definition 4.2: (Electrical Events)</u> In electrical analysis, a component is "latent" if it satisfies one of the latency conditions given by Eqs. (4.6-4.9). Otherwise, it is an "active" component making a transition from one electrical value (or state) to another. Active components generate electrical events each time they make a transition to a new value. ∎

The usefulness of this definition is seen in the following. Consider the two-stage inverter of Fig. 4.7. For this circuit, $A \in$ **Fanout(I)** and $B \in$ **Fanout(A)**. As depicted by the arcs in the corresponding graph, there are four ways to schedule nodes:

> (1) node I can schedule node A (fanout scheduling)
> (2) node A can schedule node A (self-scheduling)
> (3) node A can schedule node B (fanout scheduling)
> (4) node B can schedule node B (self-scheduling).

Whether a given node (say, node A) should actually schedule any events depends on its own state and the state of its fanouts (node B in this case). Since each node can be either "active" or "latent," a total of four cases exist. These cases are listed in Table 4.1 along with the recommended action to be taken by node A for each case.

As the table indicates, case (2) is the only case where the scheduling mechanism is conservative. The other cases do not introduce any additional work or create accuracy problems and therefore are listed as *reasonable*. However, case (2) can be a source of either accuracy problems or excessive computation. To see this, consider the circuit in Fig. 4.8. If node A is "active," it will force nodes B, C and D to be processed if the action recommended in Table 4.1 is taken. In reality, only node B should be processed. The other two nodes do not change due to the bias conditions, but this is not known *a priori*. Therefore, case (2) is considered to be a conservative scheduling strategy. The alternative would be to ask the question: is fanout x_j *sensitive* to changes in x_i? Here, $x_i = A$ and $x_j \in \textbf{Fanout}(x_i) = \{ B , C , D \}$. Only an affirmative response to this question causes a particular x_j to be scheduled by x_i.

case	status of node A	status of node B	action by node A	comment
(1)	active	active	schedule self at t+h schedule fanouts at t	reasonable
(2)	active	latent	schedule self at t+h schedule fanouts at t	conservative
(3)	latent	active	no scheduling req'd	reasonable
(4)	latent	latent	no scheduling req'd	reasonable

Table 4.1: Four Cases in Electrical Event Scheduling

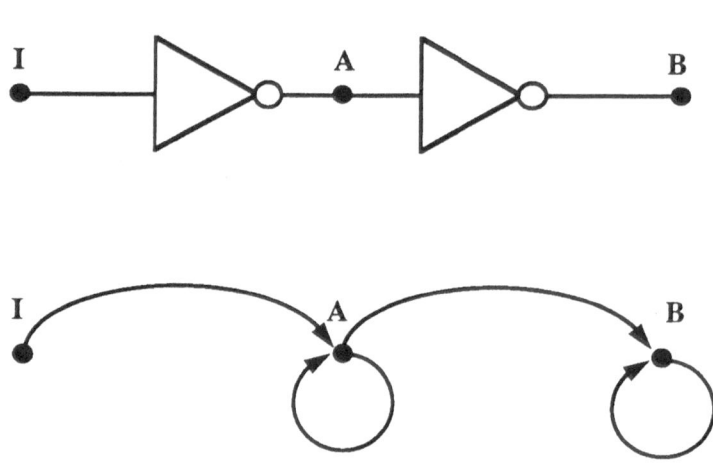

Figure 4.7: Scheduling Possibilities for a Simple Example

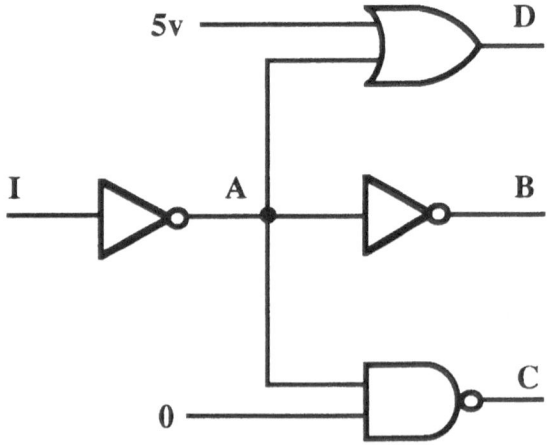

Figure 4.8: Conservative Scheduling Case

Otherwise x_j should not be scheduled.

The conditions associated with case (2) can also be viewed as a wake-up condition due to inputs. That is, "Does the change at node A wakeup node B?". The previous wake-up conditions were all handled via the self-scheduling mechanism. In this case, the question is whether or not a change at x_i translates to a change at a fanout x_j such that x_j violates its latency condition. Since x_j may have a number of fanin nodes which are active, superposition must be used to determine the combined effect of all active fanin nodes on x_j. This involves determining the transconductance, $\dfrac{\partial f_j}{\partial x_i}$, and performing the computation:

$$\Delta x = \frac{h_n}{C_j} \sum_{i=1}^{k} \frac{\partial f_j}{\partial x_i} \Delta x_i \qquad (4.16)$$

where k is the number of fanin nodes of x_j which are active, h_n is the current step size, and C_j is the total capacitance at node x_j. This computation assumes that all the additional currents, due to changes in the fanin nodes, charges the capacitances at node x_j. This produces a new wake-up condition due to the inputs, as follows:

Wake-up Condition 2:

$$h_{new} \frac{|x_{n+1} - x_n|}{h_n} + \Delta x > \varepsilon_x$$

where $h_{new} = t_{new} - t_{latent}$, and t_{new} is the current time point. In the worst-case, the computation in Eq. (4.16) can be as expensive as performing an evaluation of x_j, but it certainly is not as accurate. Since there is no way to guarantee that Wake-up Condition 2 is a sufficient check for latency violation, since it is only a local criterion, it is better to perform the evaluation of x_j rather than the sensitivity check to guarantee that an error is not made inadvertently. This results in a stronger condition for latency, which involves the fanin nodes also being latent.

The ideas presented above are formalized in the following:

(1) A component x_i is defined as being latent if

      (a) it satisfies the latency conditions specified
      in Eqs. (4.6-4.9) and

      (b) all $e_k \in$ **Fanin**(x_i) satisfy
      their latency criteria.

(2) A latent component does not generate any events.

(3) If a component is not latent, then it is active and hence will generate events for itself and for all $x_j \in$ **Fanout**(x_i) after every transition.

(4) A latent component x_i is scheduled for re-evaluation if

      (a) the wake-up condition specified in (4.10) is satisfied, or

      (b) any component $e_k \in$ **Fanin**(x_i) becomes active.

4.5.3. Latency in the Iteration Domain

Another form of latency can be exploited at each time point due to the decoupled nature of the relaxation process. Since the components in the system are changing at different rates, it is quite possible that slowly varying components will converge quickly at each time point since their behavior can be predicted accurately. Once these components have converged, there is no need to reprocess them at the same time point unless required to do so by some other component. This form of latency is called iteration domain latency and can also be exploited efficiently using the same event-driven techniques used for time domain latency.

The iteration domain is a discrete space in which a sequence of iteration values of a component can be represented as a function of the iteration number [KLE84]. This iteration domain can be viewed in the same way as the time domain. For example, if a converging sequence of iterations for a component, x_i, is plotted against the iteration number, a waveform is produced as shown in Fig. 4.9. The detection of latency in the time domain is seen to be analogous to the detection of convergence in the iteration domain. In fact, since the "step size" is fixed in the iteration domain, the check for convergence should be similar to that for the Latency Condition 1 given earlier. This corresponds to checking if the iteration waveform is "flat enough" [KLE84] and is given as

Convergence Criteria 1:

$$|x_i^{k+1} - x_i^k| < \varepsilon$$

which is consistent with the usual check for convergence. False

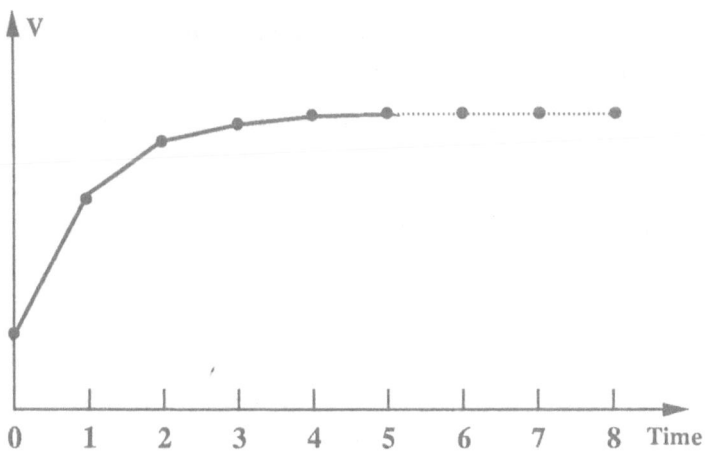

Figure 4.9: Iteration Domain Waveform

convergence occurs when the condition is satisfied but the necessary accuracy has not been obtained. Therefore, a check similar to Latency Condition 1.1 would be better to avoid this problem [KLE84].

Convergence Criteria 1.1:

$$| x_i^{k+m} - x_i^k | < \varepsilon, \quad m > 1$$

To exploit latency in the iteration domain using event-driven techniques, a table similar to the one for latency in time is necessary. In the iteration domain, if a component is "iterating," it is equivalent to being "active" in the time domain, and if it has "converged" in the iteration domain, it is equivalent to the "latent" condition in the time domain. Note that latency in time implies latency in the iteration domain, but latency in the iteration domain (i.e., convergence) *does not* imply latency in time. In fact, when a component converges in the iteration domain, a separate test is necessary to determine if it is active or latent in the time domain. The four cases in the iteration domain are listed in Table 4.2 below along with the recommended action for node A, assuming that node A is in the "converged" state initially and enters the state listed in column 2 after computing its new value.

Table 4.2 shows that case (2) is again the only conservative scheduling situation. To understand this case, consider Fig. 4.8 again. Each time node A performs an iteration, it will schedule nodes B, C and D. However, as before, only node B should be processed as nodes C and D are latent in time and hence are in the converged state at the time point. If node A requires many iterations to converge, it will schedule nodes C and D many times resulting in a lot of unnecessary work. However, there is no need to repeatedly schedule all its fanouts on every iteration, especially since the nodes have a self-scheduling ability.

Therefore, one strategy might be for node A to schedule its fanouts *on every other iteration* rather than on every iteration. This could be used for both case (1) and case (2) since the self-scheduling mechanism would take care of any additional scheduling of node B.

case	new status of node A	status of node B	action by node A	comment
(1)	iterating	iterating	schedule self at t schedule fanouts at t	reasonable
(2)	iterating	converged	schedule self at t schedule fanouts at t	conservative
(3)	converged	iterating	no scheduling req'd	reasonable
(4)	converged	converged	no scheduling req'd	reasonable

Table 4.2: Four Cases in Iteration Domain Latency

CHAPTER 5

GATE-LEVEL SIMULATION

5.1. INTRODUCTION

When the complexity of an integrated circuit design reaches the point where electrical analysis is no longer cost-effective, logic simulation or gate-level simulation may be used. Rather than dealing with voltages and currents at signal nodes, discrete logic *states* are used. In essence, logic analysis may be viewed as a simplification of timing analysis, described in the previous chapter, where the difference equations are replaced by a set of discrete state equations and only simple Boolean operations are required to obtain new logic values at each node. These Boolean operations are generally the most efficient ones available on a digital computer. In a classical logic simulator, transistors are usually grouped into logic *gates* wherever possible and modeled at the *gate-level* rather than at the individual transistor level. This form of simplification, sometimes referred to as *macromodeling,* can result in greatly enhanced execution speed by reducing both the number of models to be processed and simplifying the arithmetic operations required to process each transistor group. With event-driven, selective trace analysis and the above simplifications, asynchronous logic simulators are typically 100 to 1000 times faster than the most efficient forms of electrical analysis.

The major objective of all simulators is to accurately predict the behavior, both normal and abnormal, of the physical circuits they model. This is even more critical in the context of mixed-mode simulation where the overall accuracy may be limited by the accuracy in the higher levels of simulation. Therefore, gate level analysis in a mixed-mode simulator must provide the correct results and at least first-order timing information. The main factors controlling the accuracy of gate level simulation are the state model and the delay model. The delay model must be computationally simple and at the same time include the most important factors contributing to it. Modeling parameters are usually provided with the delay model. If these parameter values are derived from careful characterization of transistor circuits that form the logic gates, then a simplified gate model can be used with a high degree of confidence.

The tradeoff between the accuracy of logic simulation and the computer time required to perform a simulation is very important. For example, the accuracy of logic simulation can be improved by increasing the number of logic states used in the simulation. However, as the number of states increases, the overall runtime may also increase. In fact, the number of logic states, their meaning, the delay models used and the event scheduling algorithm all have a profound impact on the speed and accuracy of logic simulation. The proper choice of each of these factors depends on the circuit technology and its associated characteristics, as well as the particular design methodology used. It is this wide variety of factors that has resulted in the development of a large number of logic simulators, almost every one addressing a different set of tradeoffs.

While it is clear that the transition from the continuous electrical domain to the discrete logic domain may result in the loss of some circuit information, it is important that the circuit design methodology accommodate this type of simplification. Otherwise, the logic simulation mode cannot be used effectively. Unfortunately, in MOS logic circuits, there are many transistor configurations that are not directly amenable to this type of transformation. To overcome this problem, switch-level simulation was developed and has become the preferred form of simulation for MOS logic circuits. This approach is detailed in the next chapter.

In this chapter, some of the factors influencing the choice of logic states and delay models are described. Since logic simulators have been in use for the design of digital hardware since the early 1950s, it is impossible to address all aspects of simulator development here. Therefore, only those aspects which are related to mixed-mode simulation are emphasized. In addition, the modifications necessary to make gate level simulation suitable for the mixed-mode environment are described.

5.2. EVOLUTION OF LOGIC STATES

5.2.1. Two-State Logic Model

The earliest use of logic simulation was for the verification of combinational logic. Since the logic was assumed to have zero delay and logic gates were assumed to implement ideal Boolean operations such as AND, OR and INVERT, only two states were required: a state representing *true* (logic 1) and a state representing *false* (logic 0). With a two-state simulator, it is not only possible to verify the logic function of a digital system (i.e., generate a *truth table*) but it is also possible to detect certain other types of potential design errors such as *hazards* and *races* [EIC65]. A hazard is a momentary incorrect output state, after an input transition, resulting from paths in the circuit with different delay times to the output. There are a number of different types of hazards that can arise in in a logic circuit: static 0 hazard, static 1 hazard, dynamic 1 hazard and dynamic 0 hazard. These hazards are illustrated in Fig. 5.1. A race condition exists in an asynchronous sequential circuit if more than one of the state variables undergoes a transition during a

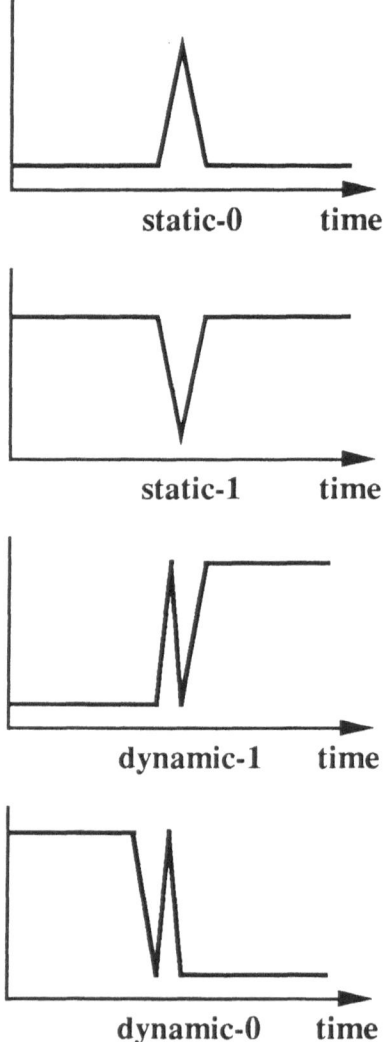

Figure 5.1: Four Types of Hazards in Logic Circuits

state transient. If the final stable state of the circuit depends on the order in which the state variables change, the race is termed *critical*; otherwise, it is termed *noncritical*.

Although hazards may occur in combinational as well as sequential circuits, they are generally most important when they affect the behavior of sequential circuits. Since hazards result from paths with different delay times, any hazard *actually causing a circuit to malfunction* will be detected as a critical race or oscillation in the circuit. However, a two-state simulator (even with random delay models) has only a limited capability for detecting races and hazards, if delay variations are not modeled. If several inputs to a logic gate change within a relatively short period of time, it is possible that the order of occurrence of these events may change if gate delays were distributed at slightly different points within their tolerance limits. If the output state of the gate depends on the order in which the inputs change, a potential hazard exists.

It is not sufficient to simply monitor the output of a gate and look for multiple transitions during an input pattern if all *potential* hazards are to be detected. Depending on the order in which the input transitions are processed, the potential hazard may or may not be detected in the zero-delay simulator. This is illustrated in Fig. 5.2 for a simple NAND gate. If input **A** changes first, then output **D** will switch to the 0 state before returning to the 1 state. However, if input **B** changes first, the output will remain at 1 during the input transitions. The potential for both static and dynamic hazards can be detected. However, the errors caused by actual circuit hazards cannot be detected in a two-state simulator without the use of more accurate delay models.

It should be noted that, in a two-state logic system, only one logic gate may drive (or fanin to) any node (often called a *net* in the context

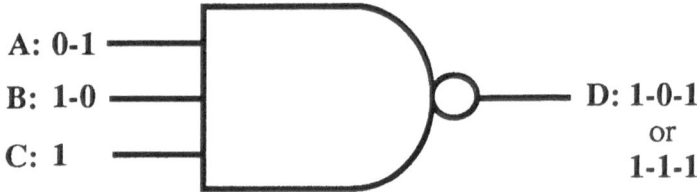

Figure 5.2: Potential Hazard in NAND Gate

of logic design). If more than one gate did fanin to a node, a potential conflict would arise if one gate had a logic 1 at its output and another a logic 0 since it would be unclear what the resulting signal at the node should be. An exceptional case is that of the wired-function (wired-AND, wired-OR), where the node is treated as a logic gate itself and performs a logic function. This is illustrated in Fig. 5.3(a) for an open-collector TTL example. If it is possible for more than one output to drive a node in a particular technology, such as the so-called *tristate* logic where gates may logically disconnect themselves from the node (as illustrated for MOS in Fig. 5.3(b)), then two-state logic analysis cannot be used to verify the design.

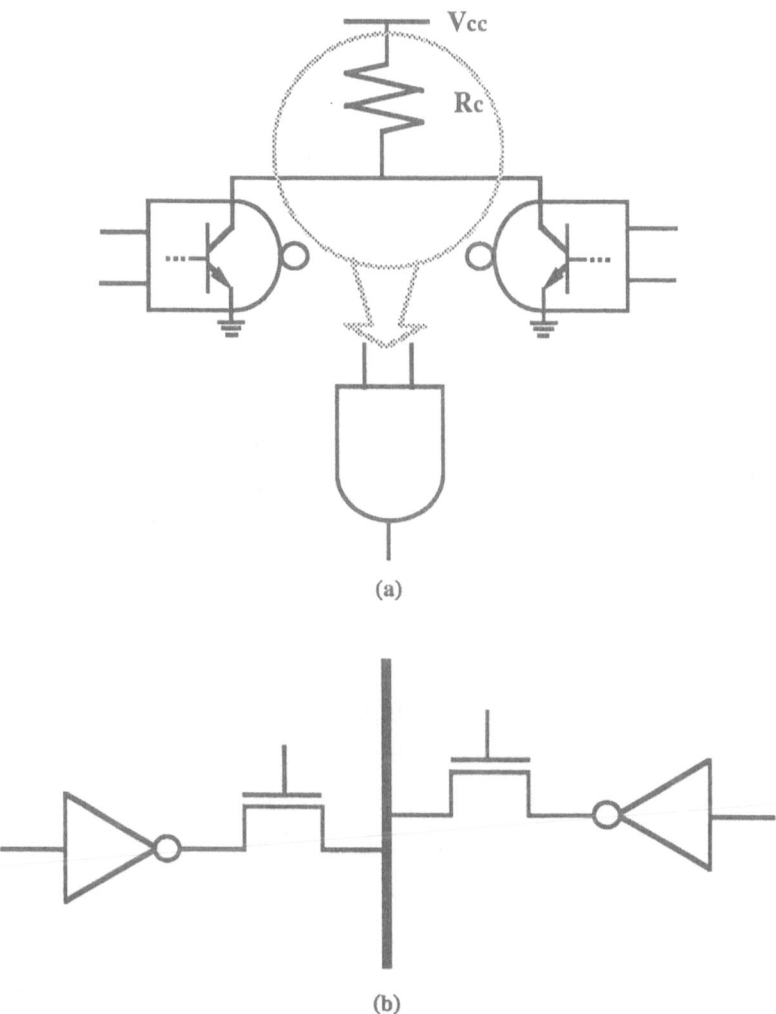

(a)

(b)

Figure 5.3: Multiple Devices Driving a Single Node
(a) Open Collector TTL Structure and Its Equivalent Logic Mode
(b) MOS Transfer Gates Connected to a Common Bus

5.2.2. Ternary Logic Model

Two-state simulation has a number of limitations. For example, if two gates drive the same node in the circuit and the output of each gate is different, a conflict situation arises. To model this conflict condition, a third state may be added--the unknown state, X. The output node is set to this X state whenever any such conflicts arise. The X state can then propagate through the fanout gates to other nodes in the circuit and possibly set them to the X state. The logic operations for the AND,. OR and INVERT gates with X gates are shown in Fig. 5.4

The simple step of adding this new state has caused much confusion and increased the complexity of logic simulation. In [BRE72], the basic problems associated with unknowns in gate-level simulation are described. One such problem arises due to the pessimistic propagation of unknowns when the value of a node is actually known. For example, in Fig. 5.5, one of the inputs is unknown, and this produces an X at each intermediate node and results in an X at the output node C. However, since a value of 1 or 0 at that input produces the same results at node C, the value at node C is actually known to be 1. Therefore, the propagation of X blindly can lead to pessimistic results and excessive computation. This problem can be resolved by keeping track of X and $\overline{X}$ values[1] during the simulation and combining them using the identities $X \cdot \overline{X} = 0$ and $X + \overline{X} = 1$ whenever they appear at common AND or OR gate inputs. A second problem with the use of the X state is due to the additional complexity it introduces into gate-level logic simulation. In fact, computing the output states of a sequential circuit with n inputs and m internal states having k out of the $n+m$ nodes unknown has been shown to be NP-complete with respect to k [CHA87].

---

[1] Multiple X and $\overline{X}$ states must be maintained, one for each different source of the X state.

a	NOT a
0	1
1	0
X	X

AND	0	X	1
0	0	0	0
X	0	X	X
1	0	X	1

OR	0	X	1
0	0	X	1
X	X	X	1
1	1	1	1

Figure 5.4: Logic Truth Tables Including X State

A number of other problems with the **X** state also exist, mainly due to the misuse of the definition of the state. For example, gate outputs must be correctly initialized prior to the analysis to either the 0 or the 1 state. If a sequential circuit is under analysis, storage nodes such as the output of flip-flops may not be known at initialization time. If the node is set to **X**, there is clearly an inconsistency since the states of Q and $\overline{Q}$ can simply be set to $Q=1$ and $\overline{Q}=0$ (or equivalently $Q=0$ and $\overline{Q}=1$) without violating the sanctity of the simulation. Consider the SR flip-flop circuit of Fig. 5.6. If the outputs are assumed to be unknown at initialization, they can not be set to known values due to the input data and the feedback of the **X** states. However, a "conflict" situation does not exist at these output nodes; therefore, the use of **X** in this case is clearly incorrect. Another state is required to account for uninitialized nodes in sequential circuits. A distinction should be made between *initial* unknowns X_1 and *generated* unknowns X_g. When an initial unknown is encountered during the simulation, it can be set to a known value in the processing of the gate it is driving. If a generated unknown is encountered, it must not be set but rather propagated through the gates. The difference between initial and generated unknown states can also prove useful in determining those parts of a circuit not exercised during the simulation (still at X_1 after the simulation).

The **X** state has also been used occasionally for the transition period between 0 and 1, which is another improper use of **X**. For this situation, a **T** state (i.e., transition state) should be employed, or possibly the **R** (rising) and **F** (falling) states to provide information about the direction of the signal transition. In mixed-mode simulation, the use of **X** is generally not recommended. However, the **R** and **F** states are extremely useful and an important part of electrically-oriented gate-level simulation, as described later in this chapter.

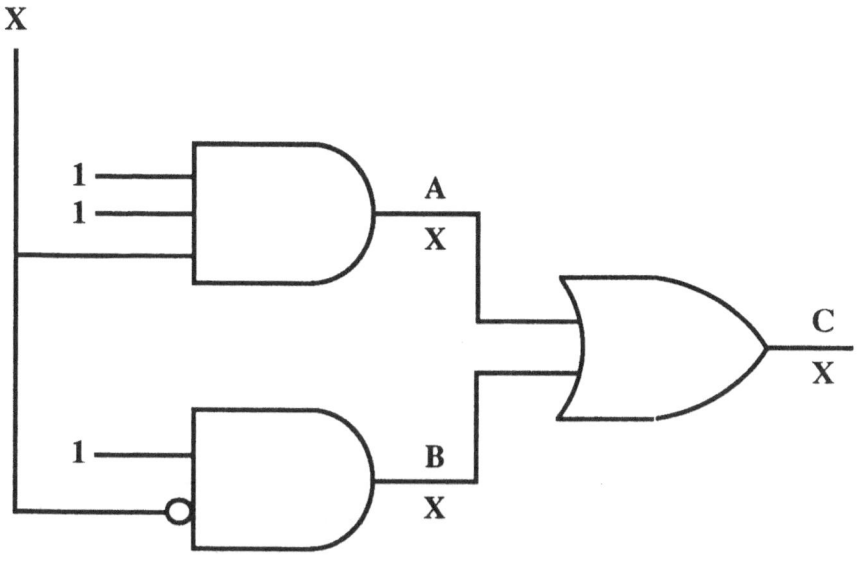

Figure 5.5: Problem Using X-State in Gate-Level Simulation

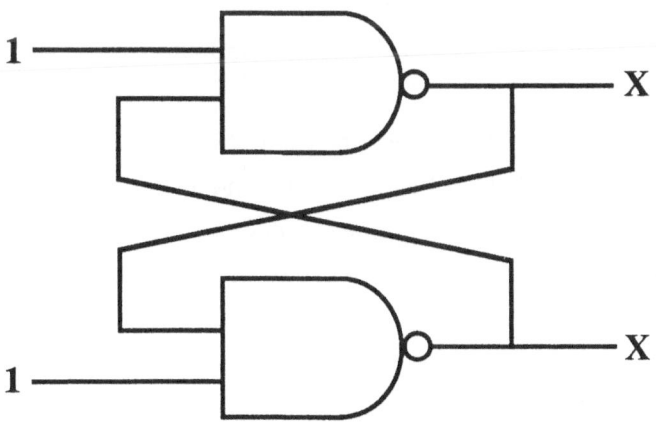

Figure 5.6: Initial Unknowns in a SR Flip-Flop

5.2.3. A Four-State Logic Model

The ternary logic model described above is still not sufficient for the analysis of general MOS digital circuits which contain transfer gates and tri-state logic circuits. For these circuits, many gate outputs may be connected to a single node, as shown in Fig. 5.7, and it is necessary to determine which output is controlling the state of the node, or *bus*. If more than one gate is forcing the node, a *bus contention* warning must be generated by the simulator. It is possible to represent the condition where the output of M1 is not controlling the bus (G1 is logic 0) by setting the output of M1 to X in that case. If this technique is used, there is no longer any distinction between the true unknown state and the *off* condition of the gate. With the addition of a fourth static state, *high impedance* (Z) or *non-forcing*, the distinction is maintained.

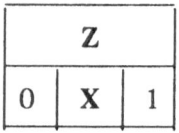

Table 5.1: Four-State Logic Simulation

The four static states are illustrated in Table 5.1. A high voltage is represented by logic 1, low voltage logic 0, and unknown is X. The fourth state, Z, is shown separately since it does not represent a voltage state but rather an impedance condition. With the addition of the Z state, bus contention can be predicted without confusion. But what if all the gates driving a node are off? What is the input to a fanout gate in this case? In MOS circuits, the previous output is generally "stored" on the parasitic capacitance at the node and held for some time. This may be modeled by saving two states at each node, the present state and the

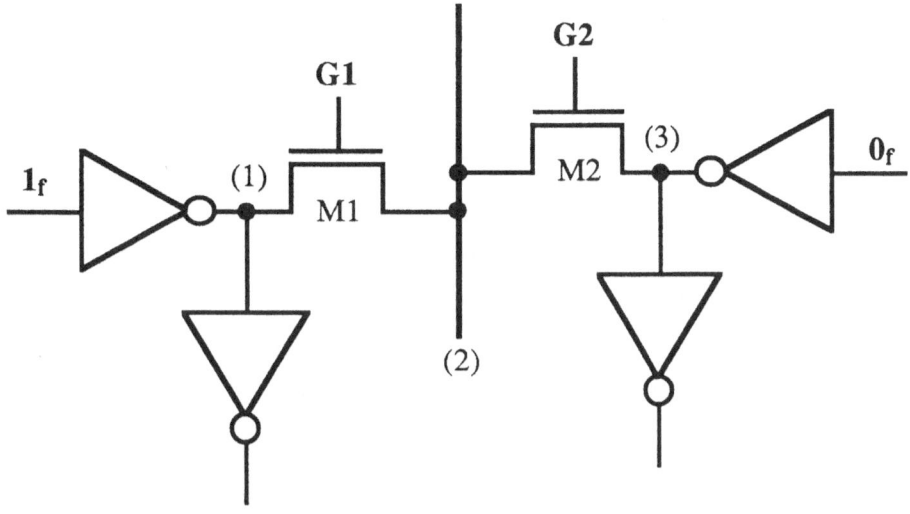

Figure 5.7: Multiple Transfer Gates at a Common Bus

previous state[2]. If the present state is **Z**, then the previous state can be used to determine the input to fanout gates.

5.2.4. A Nine-State Logic Model

Another approach that can be used to keep track of the previous state of high-impedance nodes is to add three new static states, as shown in Table 5.2. The low impedance states are called *forcing* states (0_f, X_f, 1_f), and there are now three high impedance states (0_z, X_z, and 1_z), which also carry the information about the previous signal level.

---

[2] The previous state is required to accurately model storage elements in any case.

0_z	X_z	1_z
0_f	X_f	1_f

Table 5.2: Six-State Logic Simulation

Consider once again the circuit of Fig. 5.7. If M1 and M2 are both conducting, it is clear that the state at node (2) can be determined from our simple model. But what about nodes (1) and (3)? Since the transfer gates are bidirectional, the signal at node (2) may force nodes (1) and (3) to the X state. In reality, the output impedance of the inverter is probably considerably lower than the output impedance of the transfer gate and, hence, the inverter output would determine the node state. To model this effect, another three states may be added, called *soft* states, (0_s, X_s, and 1_s), which correspond to the output of the transfer gate when its gate node is on and its input is a forcing or soft state. These states are shown in Table 5.3.

0_z	X_z	1_z
0_s	X_s	1_s
0_f	X_f	1_f

Table 5.3: Nine-State Logic Simulation

Conceptually, the y-axis of this state table may be considered an impedance axis and the x-axis as a voltage axis. In fact, the output of any logic gate may be mapped into this state table by measuring its output voltage (or current) and output impedance. As will be seen later, this technique may also be used to translate gate outputs from logic analysis into electrical inputs for mixed-mode analysis.

5.3. CHARACTERIZATION OF SWITCHING PROPERTIES

One aspect of logic simulation that takes on greater significance in the context of mixed-mode simulation is the representation of logic waveforms. In standard logic simulation, the waveforms are represented using the symbols "1" and "0" for the high and low values, respectively, and logic transitions are represented as ideal steps. The rise and fall transition times of the waveforms in standard logic simulation are not as important as the propagation delays from the input to the output of a gate. However, this is not the case in mixed-mode simulation. The transient characteristics during switching are much more important than the propagation delay. If needed, the propagation delay can always be derived from measurements on the waveforms for the input and output nodes.

It is important to have finite nonzero rise and fall delays in the mixed-mode environment for two reasons. First, from a practical viewpoint, this is not a realistic situation. The capacitance associated with each node produces some finite delay for both rising and falling signals. Second, it will undoubtedly cause convergence problems in the electrical simulation algorithms, specifically in the Newton method, due to abrupt changes in the logic waveforms that feed the electrical portions of the circuit. Therefore, the goal of logic analysis in the context of mixed-mode simulation should be to produce waveforms that are similar to the waveforms that would be generated by pure electrical simulation of the same circuit, albeit with less precision.

A modeling technique that satisfies this requirement can be developed by examining the electrical properties of gates. In Fig. 5.8(a),

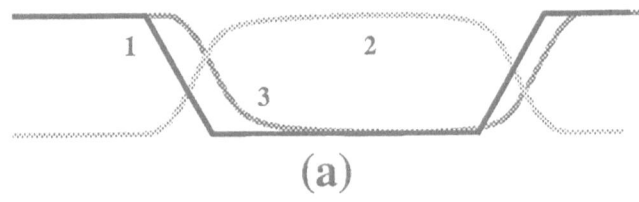

(a)

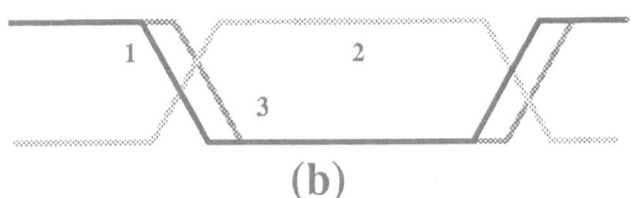

(b)

Figure 5.8: (a) Two Inverters
(b) Actual Waveforms for Inverter Chain
(c) Logic Waveforms for Inverter Chain

the output waveforms for a chain of two inverters are shown. The
waveforms are characterized by three regions: a region where the output
is low, a region where the output is high and a region where the output
is in transition. A first-order model of the charging and discharging
behaviors at each node is shown in Figs. 5.9(a) and 5.9(b), respectively.
In both cases, the model is given by an ideal current source connected to
a linear capacitor. The response at the output node is a ramp function
that is either rising or falling at a rate that depends on the value of the
capacitance and current. In reality, the charging or discharging current is
not constant so, for a first-order model, an average current must be used
to obtain the approximate timing information. In addition, the capacitance is not constant but an average can also be used for it. The logic
waveforms corresponding to the circuits in Fig. 5.9 are shown in Fig.
5.8(b). This approach can be used to generate ramp waveforms for logic
gate outputs by simply computing the rise and fall transition times. The
details of the transition time computation are left to the next section.

The three regions described above can be represented by four
parameters: a low level, a high level, a low threshold and a high threshold. These regions and parameters are shown in Fig. 5.10(a). The four
parameters have a direct correlation with the parameters that represent
the dc voltage transfer characteristic (VTC) for a logic gate as shown in
Fig. 5.10(b). This is a graph of the output voltage, V_{out}, versus the
input voltage, V_{in}, for a simple inverter. The four parameters in the
figure are as follows:

V_{OL} = low output of inverter

V_{OH} = high output of inverter

V_{IL} = maximum value of input before output begins
       to drop appreciably

V_{IH} = minimum value of input before output begins

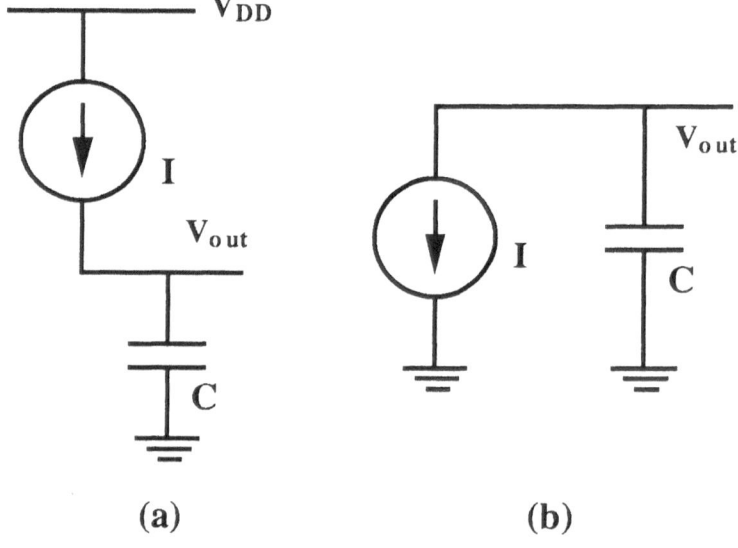

Figure 5.9: (a) First-Order Charging Model
(b) First-Order Discharging Model

to rise appreciably

These parameters are used to define the logic noise margins for the inverter:

$$NM_H = V_{OH} - V_{IL}$$

$$NM_L = V_{IL} - V_{OL}$$

The values of V_{IH} and V_{IL} based on the definitions above are somewhat arbitrary. Physically, V_{IL} is the largest value of input voltage that still maintains a valid high voltage at the output, and V_{IH} is the smallest value of input voltage that maintains a valid low output voltage. A more precise definition can be obtained by examining the input and output relationships. Clearly, the output voltage is some function of the input voltage:

$$V_{out} = f(V_{in})$$

If some voltage noise, V_{noise}, is superimposed on the input, then

$$V_{out}^{new} = f(V_{in} + V_{noise})$$

If the right-hand side is expanded in a Taylor series, then the following is obtained:

$$V_{out}^{new} = f(V_{in}) + \frac{\partial f(V_{in})}{\partial V_{in}} V_{noise} + \text{higher-order terms}$$

Therefore,

$$V_{out}^{new} = V_{out}^{old} + \text{gain} \times V_{noise} + \text{higher-order terms}$$

From this equation, assuming that the higher-order terms are negligible, it is seen that if the gain is small, the noise is attenuated. However, if the gain is large, the noise is amplified and added to the output. A

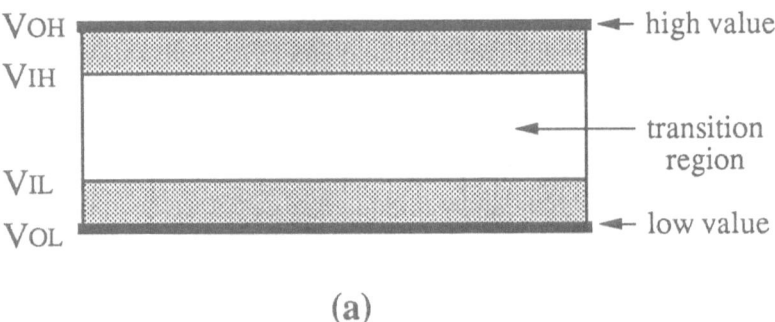

(a)

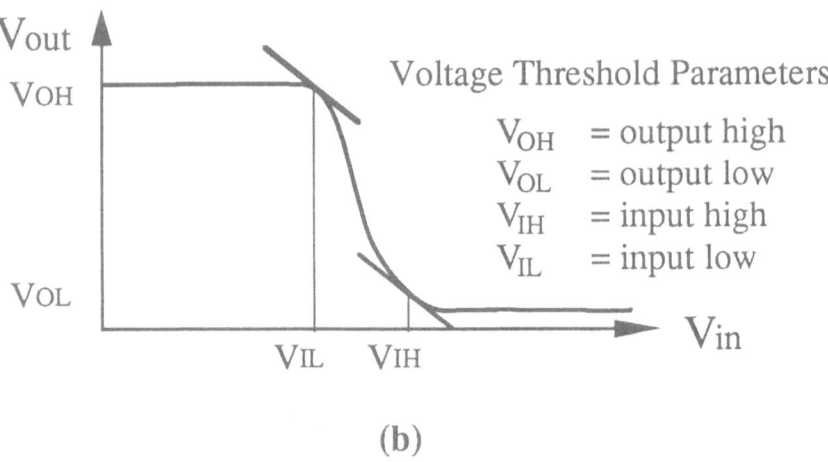

(b)

Figure 5.10: (a) Switching Regions
(b) Inverter Voltage Transfer Characteristic

reasonable breakpoint between the two cases occurs when the gain is 1. Therefore, useful definitions for both V_{IL} and V_{IH} are the points along the VTC where

$$\left| \frac{\partial V_{OUT}}{\partial V_{IN}} \right| = 1.$$

Although, in reality, the output begins to change before these two critical points are reached at the input, an ideal logic model assumes that no change will occur at the output until the thresholds are exceeded.

In terms of a logic state model, a new four-state logic model [SAK81] is needed, where the state, $s(t)$, at any node at time t is an element of the set $\{ 0, R, F, 1 \}$, where R=rising waveform and F=falling waveform. Clearly, each of the states, $s(t)$, may be defined in terms of the corresponding node voltages, $v(t)$, and the following noise margin parameters:

$$s(t) = 0 \text{ iff } v(t) \in [V_{OL}, V_{IL})$$

$$s(t) = 1 \text{ iff } v(t) \in (V_{IH}, V_{OH}]$$

$$s(t) = R \text{ iff } v(t) \in [V_{IL}, V_{IH}] \text{ and } \dot{v}(t) > 0$$

$$s(t) = F \text{ iff } v(t) \in [V_{IL}, V_{IH}] \text{ and } \dot{v}(t) < 0$$

The four-state logic model can be represented in truth table form for the AND, OR and INVERT gates as shown in Fig. 5.11. However, the actual transitions from one state to another are governed by practical considerations. Specifically, the transitions $0 \rightarrow R$, $R \rightarrow 1$, $1 \rightarrow F$, $F \rightarrow 0$, $R \rightarrow F$ and $F \rightarrow R$ are permitted. These legal state transitions can be defined in terms of a state diagram as shown in Fig. 5.12. The transitions $0 \rightarrow 1$, $1 \rightarrow 0$, $1 \rightarrow R$ and $0 \rightarrow F$ are considered to be illegal since it is

a	NOT a
L	H
F	R
R	F
H	L

AND	L	F	R	H
L	L	L	L	L
F	L	F	F	F
R	L	F	R	R
H	L	F	R	H

OR	L	F	R	H
L	L	F	R	H
F	F	F	R	H
R	R	R	R	H
H	H	H	H	H

Figure 5.11: Truth Tables for Four-State Logic Model

physically impossible to make these transitions without either visiting the intermediate states or violating the voltage limits of the circuits.

Encountering an illegal state transition during the simulation is an indication that a timing error may be present in the circuit. As an example, consider the AND gate in Fig. 5.13(a). The transitions at the inputs and outputs is specified using a string of values that indicate the state of the node in each time slot. If the two input transitions are separated in time, as in Fig. 5.13(b), there is no transition at the output. However, if the input transitions overlap, then the output may attempt an illegal state transition, which indicates that a race condition exists at the input. If input B makes the first transition but the two input transitions still

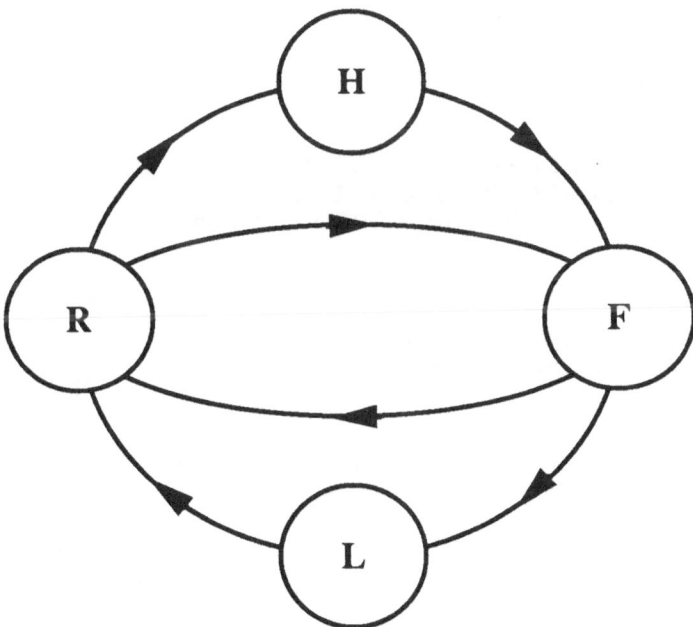

Figure 5.12: Four-State Logic Transition Model

overlap, the output will be a glitch but will not encounter any illegal states as shown in Fig. 5.13(c). Therefore, the output state transitions will either imply an error explicitly or implicitly, but in both cases a timing error can be uncovered.

5.4. LOGIC SWITCHING DELAY MODELS

Now that an appropriate logic transition model has been defined, the next step is to specify the details of the delay calculations. A variety

(a)

(b) A:11111FFFFFF000000000
    B:00000000000RRRRR1111
    C:000000000000000000000 (no change)

(c) A:11111FFFFFF000000000
    B:00000000RRRRR1111111
    C:00000000FFF000000000 (illegal transition: race)

(d) A:11111111FFFFF0000000
    B:0000RRRRRR1111111111
    C:0000RRRRFFFFF0000000 (legal transition: glitch)

Figure 5.13: Potential Timing Errors Due to Input Variations

of different delay models have been used in logic simulators and they have evolved over time due to changes in technology in much the same way as the logic model. The simplest delay model is the *zero-delay* model mentioned earlier. This type of model only allows for functional verification of logic circuits but does not allow the detection of races or hazards and, of course, it does not provide any timing information. It is also prone to problems such as "infinite looping" if there is an odd number of signal inversions in any logic feedback path. Early logic simulators used *unit delay* models to represent timing. In a unit delay simulator, all gates have the same (unit) delay between signal transitions. For logic circuits constructed from a single gate type that has similar rise and fall delays, the unit delay simulator can provide a useful analysis and lends itself to efficient implementation. If more than one gate type is used, *assignable delays* can provide improved accuracy in the results. In the assignable delay simulator, the delay of the logic gates may be assigned an integer value, T_D. This delay is a multiple of some fundamental analysis time-step, or *minimum resolvable time* (mrt). Here, the **mrt** is the minimum non-zero delay of a logic gate and its value depends on the technology being simulated. For example, the **mrt** may be 1 ns for NMOS, while a value of 100 ps may be appropriate for ECL circuits.

There are two ways in which gate delays may be interpreted as illustrated in Fig. 5.14. A *transmission line* or group delay model propagates the input patterns directly to the output, delayed by some amount T_D. Even very short pulses are propagated unaltered, as shown in Fig. 5.14(a). A second approach is to use an *inertial delay* model, in which the "inertia" or response time of the gate is modeled. If an input event occurs in less time than the time required for the gate to respond, it will be lost as shown in Fig. 5.14(b). Note that in this case *spikes* or *glitches* may be generated at the output. A spike is defined as a signal of shorter

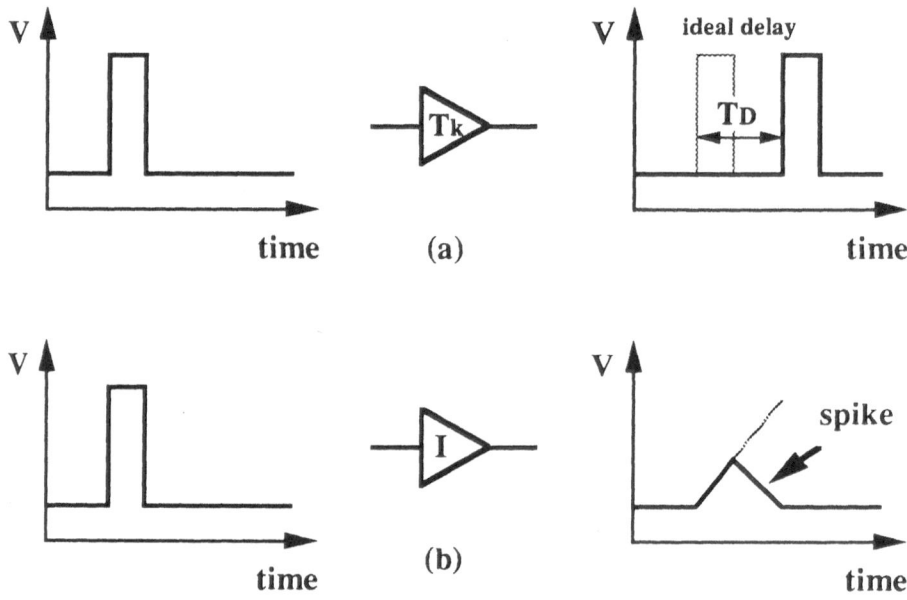

Figure 5.14: Interpretation of Gate Delay
(a) transmission line model
(b) inertial delay model

duration than necessary to change the state of an element. Spikes may
be generated by input hazards or by very narrow input pulses to a gate.
A spike may be propagated into the fanout gates as either a new state (E
for "error condition") or it may be deleted from the output and a warning message printed. The latter technique generally provides more information from the analysis since the spike is generally an error and will be
removed by the designer. By not propagating the spike, more

information may be obtained about the correct operation of the circuitry.

For mixed-mode simulation, the delay model used for the switching behavior must be derived from the electrical characteristics. The delay calculation should be based on a transition delay because of the nature of the logic model described in the previous section. For logic circuits in which rise and fall delays vary widely (such as single channel MOS), it is necessary to provide both rise (t_{LH}) and fall (t_{HL}) transition delays for each gate. These delays are a function of a number of different parameters. In MOS circuits, the switching time may depend on

- the device sizes
- the supply voltage
- the output capacitance
- the number of inputs to the gate,
  and which one switches in value
- and the shape (rise or fall times) of input waveforms.

Very few logic simulators have actually incorporated all of the above factors into the delay calculation. However, it is essential that an electrically-oriented logic simulator include the important first-order effects in the delay equation. To derive such an equation, consider the rise and fall delays of the CMOS inverter shown in Fig. 5.15. The fall time, t_{HL}, is given by [UYE88]:

$$t_{HL} = \frac{C_{out}}{\beta_N(V_1 - V_{TN})} \left\{ \frac{2V_{TN}}{V_1 - V_{TN})} + \ln\left[\frac{2(V_1 - V_{TN})}{V_0} - 1 \right] \right\}$$

and the rise time, t_{LH}, is given by [UYE88]:

$$t_{LH} = \frac{C_{out}}{\beta_P(V_1 - |V_{TP}|)} \left\{ \frac{2|V_{TP}|}{(V_1 - |V_{TP}|)} + \ln\left[\frac{2(V_1 - |V_{TP}|)}{V_0} - 1 \right] \right\}$$

where C_{out} is the loading capacitance, V_{TN} is the n-channel threshold voltage, V_{TP} is the p-channel threshold voltage, and

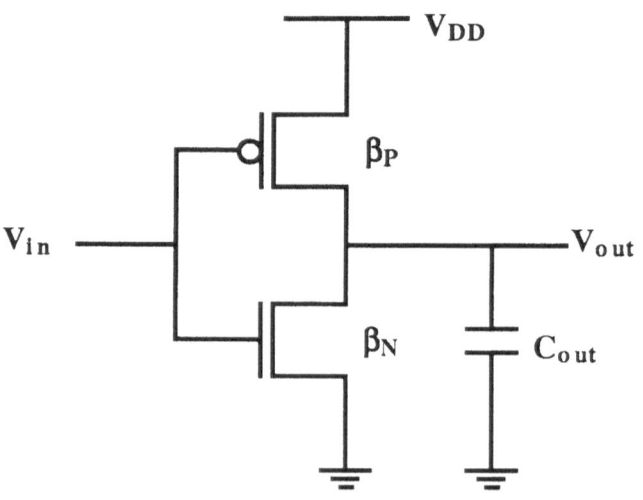

Figure 5.15: CMOS Inverter

$V_1 = V_{OH} - 0.1(V_{OH} - V_{OL})$ and $V_0 = V_{OL} + 0.1(V_{OH} - V_{OL})$ are the 90% and 10% switching points, respectively. All of these parameters are constant except for the output loading capacitance which depends on the number of fanouts connected to the output node.

The delay can be separated into two components by dividing C_{out} into $C_{intrinsic} + C_{fanout}$, where $C_{intrinsic}$ is the unloaded output capacitance and C_{fanout} is due only to external gates connected to the node. Then, the total gate delay can be represented by four parameters : the intrinsic gate delays (**tr, tf**) and the gate drive-capabilities (**trc, tfc**), where

$$\text{tr} = \text{rise time for unloaded gate (y-intercept)}$$

tf = fall time for unloaded gate (y-intercept)

trc = gate drive-capability for rising signals (slope)

tf = gate drive-capability for falling signals (slope).

Using these values, the total delays are calculated with the equations:

$$t_{LH} = tr + trc*C_{fanout} \qquad (5.1a)$$

$$t_{HL} = tf + tfc*C_{fanout} \qquad (5.1b)$$

The logic gates can be characterized to determine the four parameters (**tr, tf, trc, tfc**). The value of C_{fanout} requires that the input and output capacitances be specified for each gate. For example, **ci** can be defined as the MOS capacitance associated with the gate input and **co** can be defined as the wiring capacitance. Then the total capacitance at each node becomes

$$C_{fanout} = ci + \sum_{k=1}^{n} co_k$$

This process is shown in Fig. 5.16. Often the delay is a function of the input slope, s_i. This aspect can be incorporated into the premultipliers **trc** and **tfc**:

$$t_{LH} = tr + trc(s_i)*C_{fanout} \qquad (5.2a)$$

$$t_{HL} = tf + tfc(s_i)*C_{fanout} \qquad (5.2b)$$

It is now possible to model the delay by generating a set of curves similar to Fig. 5.17 for every primitive element (NANDs, NORs, inverters, etc.) using accurate electrical simulation. In this figure, the rise and fall times are plotted as a function of the output capacitance. A step voltage is assumed at the input of each gate. Although not strictly true, the relationship between the capacitance and delay is usually taken to be linear. That is, the delay is calculated based on the model of a constant current source charging a linear capacitor. The y-intercept of each curve represents the intrinsic unloaded rise/fall delay while the slope of each curve represents the gate "pull-up" or "pull-down" resistance. Separate characteristics are required for rising and falling outputs if the delay times are not symmetric.

5.5. LOGIC SIMULATION ALGORITHM

The following pseudo-code provides a simplified description of the logic simulation algorithm based on the previous sections. First, the event time t_n is established by *NextEventTime()*. All input sources, e_k, that are changing at that time schedule their immediate fanouts. Then all the nodes scheduled at t_n are processed in sequence until the events at that time are exhausted.

In the algorithm, a node is processed by computing the output value of its associated gate using the states of the inputs at t_n. The input state consists of a voltage value and information indicating whether the signal is rising, falling or stationary. If the new output state is different from the old one, a delay calculation is performed. If the event occurs before a transition in the opposite direction is completed, a glitch warning is produced and the original transition event is cancelled. If the event does not cause a glitch, the schedule time for each fanout of the output node, Δt_j, is computed and then the fanout is scheduled at $t_n + \Delta t$.

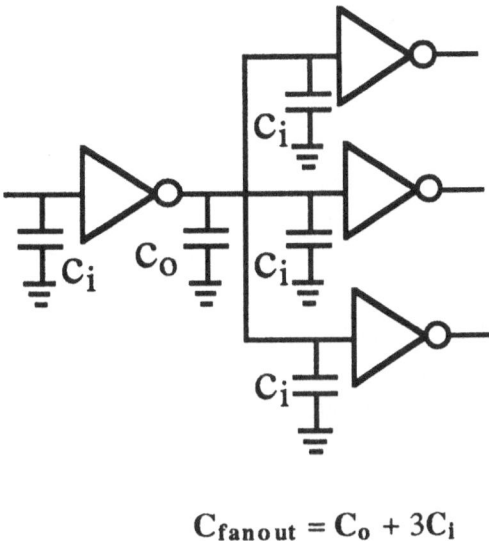

$$C_{fanout} = C_o + 3C_i$$

Figure 5.16: Computing the Total Node Capacitance

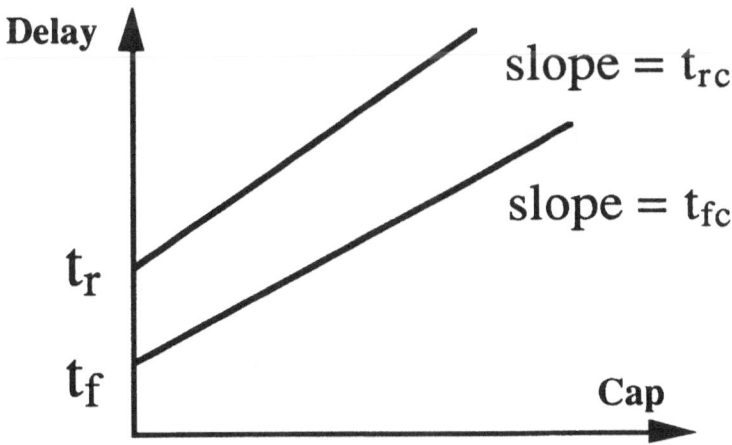

Figure 5.17: Delay vs. Capacitance for an Inverter

<u>Algorithm 5.1 (Logic Simulation Algorithm)</u>

```
    t_n ← 0;
    while (t_n ≤ T_stop) {
        t_n ← NextEventTime( t_n );
        foreach ( input k at t_n )
            if ( e_k is active )
                forall ( j ∈ Fanout(e_i) ) schedule( node j, t_n );
        /* processing logic block i */
        foreach ( node i at t_n ) {
            get input_states;
            compute new_output;
            if (node i has changed ) {
                compute delay, Δt;
                if ( current_time < last_event_time(i)) {
                    issue glitch message;
                    cancel pending events;
                }
                else { /* normal event, so schedule fanouts */
                    forall ( node k ∈ Fanout(i) ) {
                        determine schedule threshold;
                        compute schedule time Δt_k;
                        schedule ( node k , t_n+ Δt_k );
                    }
                }
            } else { /* do nothing (latency exploitation) */  }
    }
    ∎
```

CHAPTER 6

SWITCH-LEVEL TIMING SIMULATION

6.1. INTRODUCTION

Most modern logic simulators handle the problems specific to MOS integrated circuits by including the notion of signal strength in the logic model. However, the use of strength does not, by itself, solve all the modeling problems inherent to MOS circuits. For example, circuit designers use many combinations of transistors which do not have a direct mapping to a logic gate and therefore cannot be represented conveniently at the gate level. It is also difficult to model the logic operation of dynamic circuits in a convenient form in a standard logic simulator. Transfer gates further complicate the situation because they introduce dynamic loading effects, bidirectional signal flow, and capacitive charge-sharing effects. Many of these problems were resolved with the advent of the switch-level modeling and simulation technique [BRY80].

This chapter begins with a description of standard switch-level simulation and identifies a number of limitations in the approach, primarily the lack of accurate timing information, and also the fact that intermediate voltage states are not represented which may occasionally lead to incorrect results. An electrically-oriented switch-level modeling technique that resolves these and other problems is described. This technique allows variable precision simulation, thereby allowing the user to choose anywhere from logic simulation accuracy to electrical simulation accuracy. Hence, the approach effectively spans the "gap" between logic and electrical simulations. A number of other simulation approaches with similar properties are also described. In the last section, the use of this variable precision modeling approach to map signals across the interface between logic gates and electrical circuitry is described.

6.2. SWITCH-LEVEL SIMULATION

A switch-level simulator transforms an MOS transistor network into a corresponding network of switches and performs logic simulation on the resulting network. For example, in MOSSIM [BRY80], the logic circuit is described entirely at the transistor level, and the transistors are modeled as simple gate-controlled switches. The switch-level logic state model includes three logic levels (0, X, 1) and a number of strengths, s, which lie in the range $\{1, \cdots ,w\}$. Two subranges of strengths are defined, one representing all signal strengths originating at some external source, in the range $k < s < w$, and the other corresponding to nodal capacitance values, in the range $1 \leq s \leq k$. The maximum possible strength, w, is reserved for inputs only. The switch-level model attempts to incorporate the key aspects of MOS logic circuits that determine its behavior and abstract away the details of the electrical behavior. This approach greatly simplifies the algorithms needed to correctly simulate a large variety of MOS logic circuits.

The simulation process in switch-level simulation proceeds as follows. First, as a preprocessing operation, the switch-level network is partitioned into a number of subnetworks which are collections of *strongly-connected components* (SCC) or *channel-connected components*. These are sets of transistors that are connected to one another at the source or drain terminals. The identification of SCCs can also be done dynamically during the simulation process. Processing a given SCC may require a complicated series of steps, possibly involving iterations, to account for the interactions of different strengths of two or more "ON" transistors, as described below. However, the interaction between two

SCCs is easier to analyze since they are connected at the gate inputs of transistors and, hence, the logic operations do not depend on the signal strengths. The SCCs are simply scheduled and processed in the manner described earlier for logic gates using event-driven, selective trace techniques. Therefore, this mode of simulation well-suited to implementation in mixed-mode simulators.

The complicating factor in the processing of SCCs is due to the bidirectionality of transfer gates, or pass transistors. Although the transfer gate is inherently a bidirectional element, it is usually found in applications in which the signal flow is intended to be unidirectional. That is, the circuit designer expects signals to flow in only one direction through the device. However, there are occasions when transfer gates are used in bidirectional applications, or other situations in which a design error leads to signal flow in different directions at different times. A simulator must be able to analyze these cases accurately if it is to be useful. There have been a variety of modeling approaches for bidirectional transfer gates, including the unconventional approach of two unidirectional elements connected back-to-back. This approach can lead to inconsistencies when different logic values are on opposite sides of the element. Each value can flow through one of the transfer gates and reach the opposite side and then propagate through the circuit producing incorrect results.

During the evaluation of an internal node of an SCC, the elements connected to that node try to impose their values on the node and the final state is determined by the element with the highest strength. A path analysis is actually performed to identify all possible paths from the node to a supply or ground node [BRY84]. Each path is assigned a strength that depends on the transistor with the lowest strength. Weak paths are blocked at intermediate nodes if a stronger path is encountered at the nodes. The strongest path to a given node determines the final state of the node. If two paths of equal strength but opposite values are encountered at a given node, the node is assigned to the **X** state.

The path analysis approach is effective for handling bidirectional signal flow. A simpler approach is to use the so-called "supernode" technique [BRY80]. In this approach, all nodes that are connected through transistors that are "ON" are considered to be the same node for processing purposes. All devices connected to this composite node are processed together to determine the new state. The new state is then assigned to all nodes which comprise the supernode. The main problem with this approach is that it cannot adequately handle the case where the final values at the nodes are different and determined by the strengths of the transistors in the subnetwork. Therefore, it does not permit different nodes of a supernode to reach different values.

An alternative to the supernode and path analysis approaches is to use an iterative or relaxation-based method to determine the new states of these strongly-connected nodes (SCN) [DUM86]. The first step in this approach is to assign all nodes to the lowest strength permissible, or to a strength associated with the capacitance at each node. The signals are then propagated from the source nodes through the switch network starting from the signal possessing the largest strength. This processing order prevents the accidental propagation of weaker signals onto storage nodes that may inadvertently generate the unknown logic level. The internal nodes of the SCC are evaluated using local event-driven techniques and the fanout nodes within the set of SCNs are scheduled whenever they change state. The process is repeated until convergence is obtained, at which point scheduling occurs at the SCC level, i.e., the fanout SCCs are scheduled. Note that by using iterative methods, the nodes within a strongly connected component may converge to different logic levels, which is the main advantage of the approach. The combination of local relaxation methods at the SCN level and standard event-driven methods at the SCC level allows efficient switch-level analysis to be performed.

One problem not addressed above is that of processing transfer gates with unknowns at gate inputs (i.e., X-transistors). The strategy in analyzing circuits with X-transistors is to minimize the number of X states generated at the internal nodes of an SCC. A pessimistic approach would be to generate the X state at each of the drain and source output nodes of each X-transistor. This method is the easiest to implement but it may actually force the simulator to process many more events than necessary since the X state tends to be "sticky" and propagates throughout the circuit very quickly [CHA87]. A brute-force approach would be to enumerate all possible combinations of gate input values by replacing the X-transistors by either 1-transistors or 0-transistors. If there are k X-transistors, the SCC would have to be evaluated a total of 2^k times! Any node which produces the same logic level, regardless of the input combination, is set to that logic level; otherwise, it is set to X.

A better approach [BRY87], which offers linear computational complexity in k, is to first choose the gate settings of the X-transistors to maximize the number of 1's or X's in the SCC under consideration. Then, the process is repeated to select the gate settings to maximize the number of 0's or X's in the SCC. Again, any node which reaches the same logic level in both cases is set to that level; otherwise, it is set to X. This approach has been shown to produce the same results as the computationally expensive method described above but requires much less work.

6.3. A GENERALIZATION OF THE NINE-STATE LOGIC MODEL

Switch-level simulation has been adopted as an efficient technique for functional verification of large MOS digital circuits. However, there are many circuit configurations that may lead to incorrect solutions when the simple switch-level model is used. In fact, simple examples can be constructed that require more than the three strengths and three states of the nine-state logic model (described in Chapter 5) to produce the correct solutions. To illustrate this point further, consider the two circuit fragments shown in Fig. 6.1. The circuit in Fig. 6.1(a) is a 2-ϕ regenerative latch driving a bus. The two inverter stages provide forward gain while the depletion load device provides a resistive feedback path from C to A. In this case, at least four strengths are necessary to obtain the correct results at node C: inverter inv2 has a weak pull-up strength W_1, a forcing pull-down strength F and a high-impedance pull-down strength H (depending on whether it is on or off). The depletion device has a resistive strength W_3, such that W_3 is less than W_1. In addition, the pass transistor connected to the bus also has a strength, W_2, which is less than W_1 but greater than W_3. As described earlier, switch-level simulators provide a range of strengths to address this problem.

A more serious problem is that the switch-level model may not produce the correct results for an arbitrary connection of pass transistions when threshold voltage drops are important. For example, the situation shown in Fig. 6.1(b) is a case where additional voltage states are necessary. Here, the designer has inadvertently connected the gate of transistor M_3 to a node which is already two threshold voltage drops below the input signal. Therefore, the value at node D can only rise to three threshold voltage drops below the input value and this may not be high enough to be considered as a valid high. While this circuit is clearly a poor design, it is important for a simulator to detect this type of

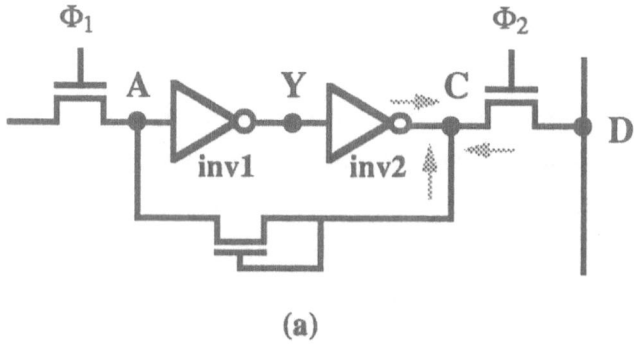

(a)

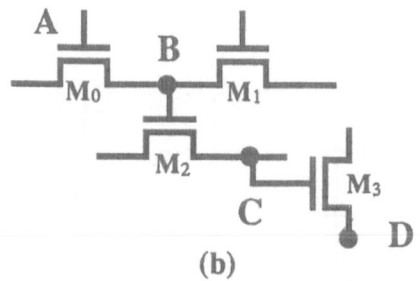

(b)

Figure 6.1: Examples Requiring Additional States and Strengths

error. A standard switch-level simulator would not be capable of identifying this error since it uses only three logic values. For this case, at least three additional logic values are required. Hence, an appropriate state model to adequately simulate both circuits in Fig. 6.1 is shown in Fig. 6.2.

Another situation which requires multiple strengths and voltage levels is the simulation of dynamic circuits. In these circuits, capacitive charge-sharing and feedthrough effects often degrade the voltage levels. Feedthrough usually occurs when a clock signal feeds through a floating capacitor to an isolated node with a grounded capacitance. The isolated node also sustains a sharp voltage transition which depends on the value

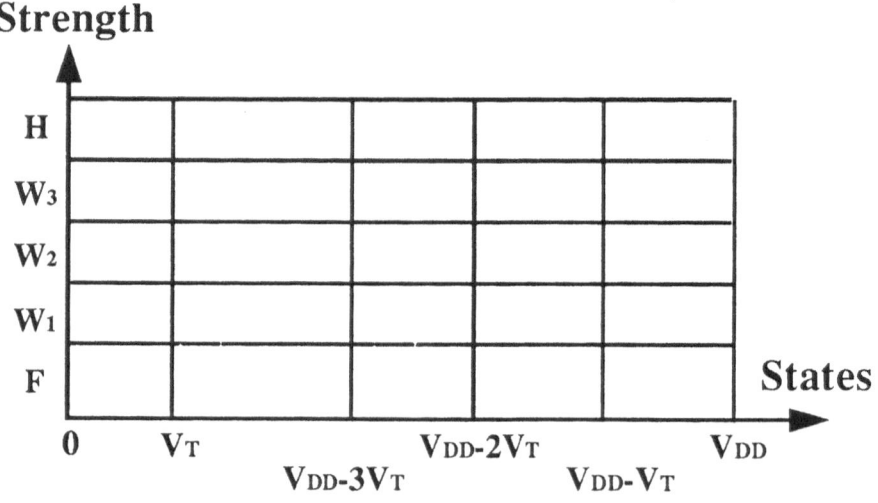

Figure 6.2: Better Logic Model for Simulation of Circuits in Fig. 6.1

of the grounded and floating capacitance values. Usually, feedthrough is not a significant factor. However, charge-sharing can often lead to circuits that do not function properly. Charge-sharing occurs when a transistor connecting two isolated grounded capacitors is turned on. The total charge is redistributed between the two capacitors until their node voltages are equal. It is possible to handle charge-sharing without introducing additional logic levels by assigning to each node a strength that corresponds to its capacitance value. If charge-sharing occurs, the node with the larger capacitance imposes its value on the node with the smaller capacitance (for worst-case analysis) and a potential problem is at least observed [BRY80].

One basic limitation of standard switch-level simulation still remains: accurate timing information is not provided. Electrical simulation provides detailed timing information but is very expensive due to the use of complex analytical models that characterize the transistor current-voltage relationships. Logic simulation is extremely fast but is often unable to provide more than first-order timing information using simple expressions to compute the rise and fall delays. Clearly there exists a large "gap" between electrical simulation and logic simulation. The arguments made above promoting multiple logic values and strengths and the requirement for switch-level timing simulation can be resolved by treating strengths as electrical resistances and logic states as electrical voltage levels. This connection allows a generalization of the model of strength vs. state in logic simulation to resistance, R, vs. voltage, V, in electrical simulation.

The $R-V$ characteristics for an inverter driving a pass transistor are shown in Fig. 6.3 based on SPICE2 simulations. These are dc transfer curves of the output resistance of the inverter and the output resistance of the transfer gate as a function of their respective output voltages. The

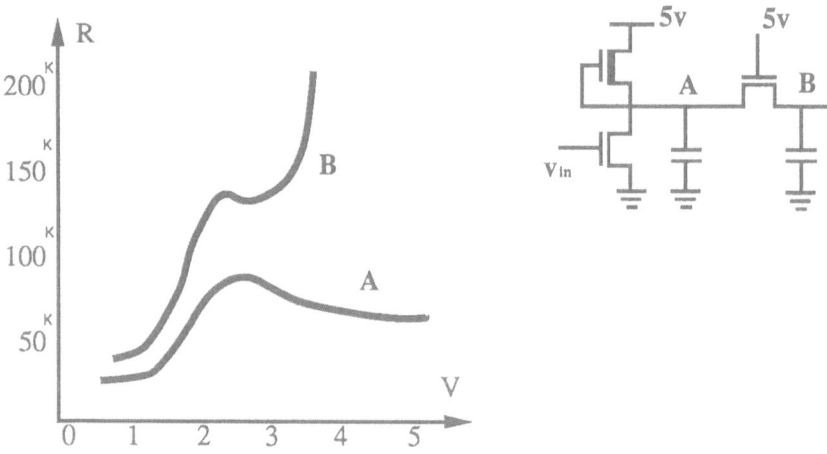

Figure 6.3: Resistance vs. Voltage Plane

two curves are highly nonlinear and do not exhibit monotonic behavior. Conceptually, electrical simulation has an infinite number of allowed "states" in this plane while the higher levels of simulation discretize the horizontal and vertical axes into a finite number of states. As a result, the difference between electrical and logic simulations becomes one of the degree of discretization of the **R**–**V** plane. This relationship also provides a convenient way of mapping from one form of simulation to the other in mixed-mode simulation. The use of this model as a vehicle for simulation is described in the next section. Its application in signal mapping across the mixed-mode interface is addressed in Section 6.6.

6.4. SIMULATION USING THE GENERALIZED MODEL

6.4.1. Electrical-Logic Simulation

A variable precision simulation approach, called *electrical-logic simulation* or simply *Elogic* [KIM84], has been developed based on the generalized model described in the previous section. This form of simulation can be viewed as a relaxation-based, switch-level simulation technique. Elogic uses electrical device models in the context of switch-level simulation which allows electrical timing information to be obtained. As part of the Elogic modeling process, a number of discrete voltage levels are selected. These levels need not be equally spaced but the number of levels and their values have an impact on performance and accuracy. In standard electrical simulators, the time-step is selected first and then the node voltage change is computed. By contrast, in Elogic the voltage step is known in advance and the *time* required to make a transition from one voltage state to another adjacent voltage state is computed. Similar approaches are used in SPECS [DEG84], MOTIS3 [TSA85], SPECS2 [VIS86] and ADEPT [ODR86], as described in the next section.

The processing sequence in Elogic is illustrated in Fig. 6.4 for a simple inverter example. The set of Elogic states is defined to be V_0, V_1, V_2, V_3, and V_4. As shown in Fig. 6.4(b), the input makes a sequence of transitions from V_0 to V_4 and visits each intermediate state between the two end points. Each transition at the input node causes an event to be scheduled at the output node. The corresponding output computed by Elogic is illustrated in Fig. 6.4(c). Note that the first transition at the input does not cause a transition at the output node since the transistor does not turn on. However, the second transition and all subsequent input transitions result in transitions at the output. Note also that the output continues to make transitions even after the input reaches

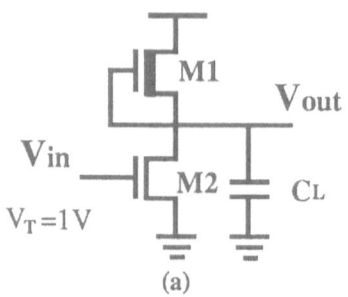

(a)

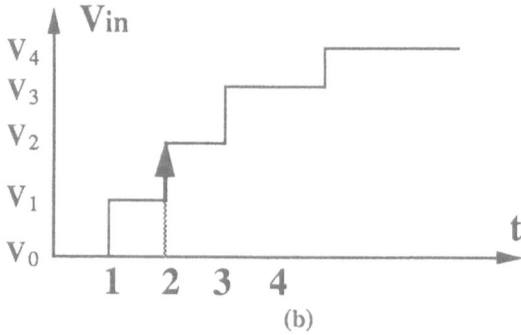

(b)

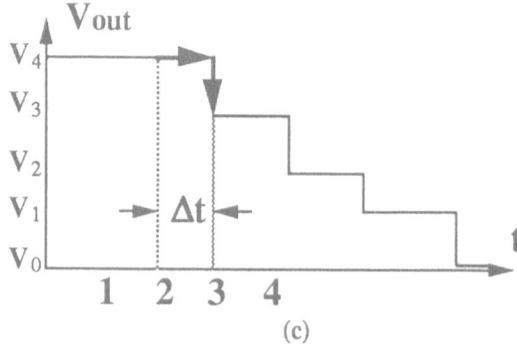

(c)

Figure 6.4: Elogic Processing Sequence for a Simple Inverter

its final value due to a self-scheduling mechansim.

Briefly, the steps required to calculate the transition time, Δt, are as follows: each nonlinear device is first replaced by a linearized equivalent model. This model is used to compute the steady-state or final voltage, V_{ss}. An exponential characteristic is used to predict the transient behavior of the voltage at the output node from the present state, V_n, to the final value, V_{ss}. The transition time, Δt, is then computed as the time required to go from V_n to V_{n+1} along this exponential characteristic. After the input has completed its sequence of transition events, the output still continues to be scheduled due to its own self-scheduling mechanism, similar to the one described for ITA in Chapter 4. The output will continue to schedule itself until it reaches the steady-state level.

The algorithm is modified slightly if the input makes a new transition *before* the output has completed its current transition. This situation is usually categorized as a glitch in logic simulation but it calls for the rescheduling of a pending event in Elogic. If the output is very close to the next state, V_{n+1}, it is set to the next state and a new event is scheduled only if a transition to V_{n+2} is warranted. If the output is still very close to the previous state, V_n, it is reset to the previous state and a new event time is calculated for the transition to V_{n+1}. If the output is somewhere in between the two states, a new transition time is calculated using V_n and the new value of the input node. The event is then rescheduled at the average of the original event time and the new event time.

The number of Elogic voltage levels selected and their position have an important impact on the accuracy and speed of simulation. Specifically, the precision with which a given voltage can be represented is limited by the set of voltage levels chosen in an Elogic model. If the actual value of a node voltage is between two Elogic states, the node voltage must be set to the closest defined level. This operation is analogous to a roundoff process and it produces a roundoff error. The number of states can always be increased to improve the precision in representing a particular voltage. However, since it is necessary to visit each intermediate state whenever a transition is made from some initial state to the final state, the simulation time increases as the number of states increases. It is this continuous tradeoff between speed and precision that makes Elogic particularly attractive as it effectively spans the large speed/precision "gap" between classical electrical and logic simulations. The user can use very few states in the preliminary design phase to verify the functionality of the circuit and obtain crude first-order timing estimates. As the design matures, more and more states can be added as necessary to improve the accuracy of the analysis. In addition, different parts of the same circuit can be simulated using a different number of states; this constitutes mixed-precision simulation, which is a special form of mixed-mode simulation.

The detailed calculations for the transition time are now described using the two nonlinear devices connected to a linear grounded capacitor given in Fig. 6.5(a). Assume that the initial state of the node is V_n. When the output node is processed, the nonlinear devices are first converted to linear devices. This can be done using either a small-signal model, which uses the incremental conductance and current of the device relative to a given operating point, or a line-through-origin model which uses the large-signal conductance of the device. In either case, the model is obtained by a table lookup scheme. The linear equivalent network following this step is shown in Fig. 6.5(b). From this circuit, it is clear that the steady-state value of the output node is

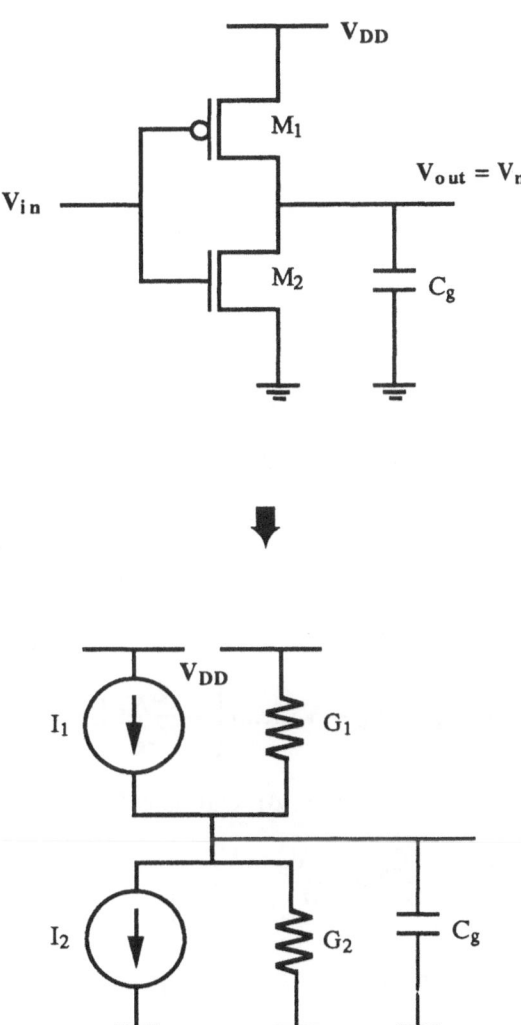

Figure 6.5: Circuits Used to Calculate Transition Time

$$V_{ss} = I_N/G_N \tag{6.1}$$

where $I_N = \sum_{i=1}^{n} I_i$ and $G_N = \sum_{i=1}^{n} G_i$. The next step is to determine if the node will undergo a transition from the present state to another adjacent state. This is done by checking if either:

$$V_{ss} > V_n \tag{6.2a}$$

or

$$V_{ss} < V_n \tag{6.2b}$$

If either condition is true, a transition time calculation is warranted.

The dynamic behavior of the linear equivalent circuit is given by

$$C_g \dot{V} = I_N - G_N V, \quad V(0)=V_n \tag{6.3}$$

for which the closed form solution is

$$V(t) = V_{ss} + (V_n - V_{ss})\exp\left[-\frac{G_N}{C_g}t\right]. \tag{6.4}$$

Using this equation, the transition time, Δt, can be calculated as follows:

$$\Delta t = \frac{C_g}{G_N}\ln\left[\frac{V_n - V_{ss}}{V_{n+1} - V_{ss}}\right] \tag{6.5}$$

A problem with this approach is that an expensive log function is required to calculate Δt every time a node is evaluated. One way to avoid this function evaluation is to use a table lookup log function. Another approach is to use a linear charging model in place of the closed-form solution. This approximation assumes that the excess current available from the current source is constant during the transition from one state to the next. In reality, the charging or discharging current for the capacitor tends to decrease as a function of time; therefore, the model is always optimistic. This model can be derived by applying the forward-Euler integration method to Eq. (6.3)

$$\frac{C_g}{\Delta t}(V_{n+1} - V_n) = I_N - G_N V_n \qquad (6.6)$$

Then,

$$\Delta t = \frac{C_g(V_{n+1} - V_n)}{I_N - G_N V_n} \qquad (6.7)$$

6.4.2. The Elogic Algorithm

The details of the Elogic simulation algorithm are presented below. In the algorithm, Elogic is implemented using event-driven techniques since a node schedules its fanouts for processing only when it achieves a new state. If a fanout node has already been scheduled at a some time t_E, in the future, it will be rescheduled at the present time, t_i.

Algorithm 6.1: (Electrical-Logic Simulation Algorithm)

```
/* processing node i */
    if ( t_i=t_{n+1} OR V_i≈V_{n+1} ) { /* reached new state */
        recompute ← FALSE;
        update voltage, V_i←V_{n+1};
        /* fanout scheduling */
        forall ( fanout nodes k of node i )
            schedule ( node k at time t_i );
    }
    else { /* did not reach new state */
        reset voltage, V_i←V_n;
        reset time, t_i←t_n;
        if ( V_i≈V_n ) recompute ← FALSE;
        else recompute ← TRUE;
```

```
        }
        G_N ← 0 and I_N ← 0;
        forall (fanin nodes k of node i) {
            replace node k by a constant voltage source;
            compute G_k and I_k;
            update G_N and I_N;
        }
        compute V_ss, the steady-state voltage;
        /* Check for transition using Eq. (6.2) */
        if ( node i can make a transition ) {
            transition ← TRUE;
            compute transition time, Δt, using Eq. (6.2) or (6.7);
            if (recompute = TRUE) Δt=(Δt_n+Δt_{n+1})/2;
        }
        else transition ← FALSE;
        if ( transition = TRUE ) {
            if ( (t_i + Δt) < T_STOP ) /* self-scheduling */
                schedule ( node i at t_i + Δt );
        }
        else {do nothing}; /* latency exploitation */
```

■

6.4.3. Problems with the Elogic Approach

The Elogic algorithm, if implemented exactly as described above, may encounter certain problems that lead to excessive computer run times or reduced accuracy. The first problem is that of algorithmic oscillation of a node voltage where one does not exist in the true solution. The simple form of this problem arises if the steady-state solution, V_{ss}, lies between two discrete Elogic states. For example, if V_{ss} lies in the range $V_1 < V_{ss} < V_2$, then the node will be assigned to the value V_1 or

V_2, whichever is closer to V_{ss}. However, if the node is re-evaluated using the new Elogic state voltage, it may force the node to move in the opposite direction, in which case it will be set to the other neighboring state. Again, since the true solution is in between the two defined states, the node will attempt to make another transition in the opposite direction creating an oscillation situation.

One approach to resolve this problem is to detect oscillations and then suppress them. This may lead to the inadvertent suppression of actual oscillations in the circuit; therefore, it is not an attractive solution. Another approach is to introduce hysteresis into the state transition criterion whenever the node voltage changes direction. Simple oscillation usually occurs as the steady-state voltage is reached. Therefore, if the sign changes on the time derivative of voltage, it is appropriate to require a significant change in the value before a transition in the opposite direction is undertaken. For example, the transition could be scheduled if the new steady-state voltage is beyond the midpoint of voltage region just visited during the last transition. Using this strategy, a transition occurs only if $V_{ss} > (V_n+V_{n+1})/2$ or $V_{ss} < (V_n+V_{n-1})/2$. Another way to resolve the problem is to simply set the node to an intermediate "illegal" voltage level when the steady-state interval is reached. The node is permitted to leave this illegal state only if it is scheduled by another node.

There is a second source of oscillation, termed *interactive oscillation*, which is more insidious and involves two or more nodes. As shown for the circuit in Fig. 6.6(a), the problem occurs when two neighboring nodes use each other's values to determine their next states and the true solution lies between two Elogic states. In this case, node A is scheduled to make a transition from 1V to 0V, while node B is scheduled to make a transition from 0V to 1V. However, after the transitions occur, both nodes make a transition in the opposite direction, and this process continues indefinitely. This type of oscillation is more difficult to detect than the simple oscillations described earlier, but the problem can be solved by introducing more states into the Elogic model.

A third problem arises due to strong coupling between two or more nodes in the circuit. This problem can be illustrated using a simple circuit as shown in Fig. 6.6(c), where $G_1=1$mho and $G_2=9$mhos, and initially $V_A = V_B = 0$ V. Note that in evaluating node A, a zero volt source is applied at node B thereby grounding it. The Norton equivalent model seen by node A is computed as follows:

$$I_N = 5\ G_1 = 5 \times 1 = 5\ \text{(A)} \tag{6.8a}$$

$$G_N = G_1 + G_2 = 1 + 9 = 10\ \text{(mhos)} \tag{6.8b}$$

Therefore, the Thevenin equivalent voltage is

$$V_{ss} = I_N\ /\ G_N = 0.5\ \text{V} \tag{6.8c}$$

Clearly, if the voltage change necessary to warrant a transition is larger than 0.5V, the basic Elogic method would not attempt to transfer node A to the next adjacent state. As a result, both V_A and V_B would remain at zero volts. As described earlier, strong coupling affects the convergence speed of ITA and WR, whereas in the case of Elogic, it results in a transition error. For this circuit, the maximum voltage step which can be used depends on the ratio of G_1 and G_2 and, in general, the Elogic states for a given problem should be selected with this rule in mind. Another solution to this problem is to determine the steady-state voltages of all nodes in a set of SCNs using switch-level techniques, and then schedule transitions based on this analysis [TSA85].

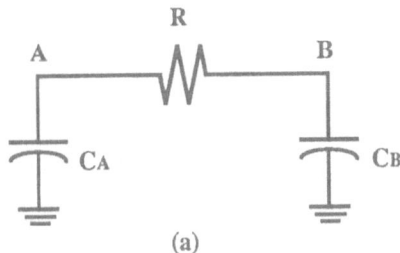

(a)

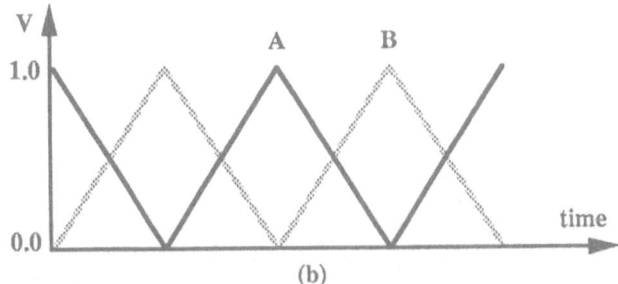

(b)

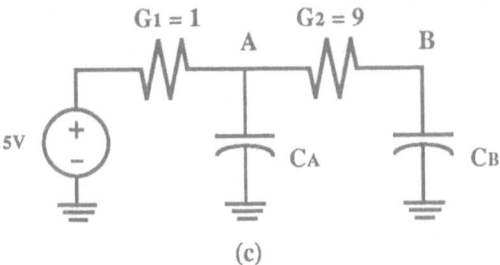

(c)

Figure 6.6: Simple Elogic Problems
(a) example circuit (b) interactive oscillation (c) strong coupling

6.5. A SURVEY OF SWITCH-LEVEL TIMING SIMULATORS

A number of other switch-level timing simulation techniques have been developed over the past decade that are also appropriate for use in a mixed-mode simulator. The original work in this area was, of course, the timing simulation algorithms of MOTIS [CHA75] as described in Chapter 4. More recently, there have been a number of notable contributions that are embodied in the programs RSIM [TER83], SPECS [DEG84], MOTIS3 [TSA85], ADEPT[ODR86], SPECS2 [VIS86], and iDSIM [RAO89]. The techniques used in these programs are reviewed briefly below.

The RSIM program attempted to produce timing waveforms for the switch-level technique by adding a linear resistor in series with each transistor switch and providing a capacitor to ground at each node. The value of the resistor was set to infinity when the gate voltage was low and to some finite resistance when the gate was high. Resistance values were calculated using the length and width of the transistors. The logic state model included only 0, X and 1. During the simulation, the transistors were replaced by their equivalent resistances and then combined to form a Thevenin equivalent circuit, with resistance R_{drive} and voltage source V_{thev}, driving a loading capacitance, C_{load}. When a transition was expected at a node, the time required to make the transition was computed as $R_{drive} \times C_{load}$. Since the values of R_{drive} for low-to-high transitions and high-to-low transitions were computed using different values of resistances for the transistor, the accuracy was often within 30% of SPICE2 for many circuits while providing over two orders of magnitude of speed improvement. The Elogic method can be viewed as an extension of this approach with the flexibility of allowing more states and having table lookup equivalent models for the transistors in each state.

The "fast timing" simulation approach of MOTIS3 is based on the Elogic algorithm. However, a backward correction scheme is used with the variable voltage step scheme to improve accuracy and avoid oscillatory behavior. First, the net current, I_{net}, available to the charge the load capacitance, C_{load}, is calculated. Then the time required to make the transition is calculated using either an exponential model or the forward-Euler model (shown here):

$$h_n = (V_{n+1} - V_n)/I_{net} \qquad (6.9)$$

Next, the value h_n is used to perform a regular integration step to compute a new target voltage, V'_{n+1}. Finally, h_n is scaled to produce the actual event time:

$$h'_n = h_n(V'_{n+1} - V_n)/(V_{n+1} - V_n) \qquad (6.10)$$

One additional contribution in MOTIS3 is the use of a so-called "superblock" approach to handle tightly-coupled nodes. First, the steady-state voltage, V_{ss}, of every node in the superblock is computed. Then, the delay calculation for each node and the minimum delay is assigned to the superblock. The node voltages in the superblock are scaled with respect to this delay.

The ADEPT approach is also similar in many ways to the basic Elogic approach described in the previous section. Like the MOTIS3 approach, it allows variable voltage steps to be used to improve accuracy at the expense of additional CPU-time. However, its most distinguishing feature is the implicit dynamic partitioning approach used to process tightly-coupled nodes. In ADEPT, when a node i is computed, the nodes, $\{j\}$, that are neighbors of i are checked for tight coupling to i using the criterion:

$$\frac{C_{ij}\dot{V}_j + G_{ij}V_j}{I_i} > \varepsilon \qquad (6.11)$$

All neighboring nodes that satisfy this criterion are solved using local relaxation methods to produce the correct results. Since this is applied to every node separately, it can be viewed as overlapped, dynamic partitioning.

Another promising variable precision approach has been implemented in the SPECS2 program, which is based in part on the techniques used in SPECS [DEG84]. A tree/link based equation formulation [CHU75] is used in the program, instead of the standard nodal formulation described in Chapter 2. This approach is well-suited to the simulation of circuits containing ideal switches that have infinite resistance when OFF and infinite conductance when ON. Devices with these properties are very troublesome in the context of nodal analysis. In tree/link based analysis, a circuit graph is constructed from the circuit description and a tree is identified in the graph. A *tree* is defined as a connected, acyclic subgraph that contains all the nodes of the original graph. The branches that belong to the graph are called *tree branches* while the remaining branches are the *links*. The links combine to form a *cotree*. Once the tree has been defined, a *cutset* is identified in the tree. A cutset is some subset of the branches of a tree such that their removal results in a graph that is no longer connected, but the insertion of any one of the branches from the removed set results in a connected graph. Cutsets are the subgraphs to which KCL is applied, and loops are the subgraphs to which KVL is applied. The fundamental cutsets and loops are used to formulate the circuit equations.

SPECS2 uses table models to define the device I-V characteristics as shown in Fig. 6.7(a). Note that the segments are piecewise constant, forming a set of step functions, as opposed to being piecewise linear.

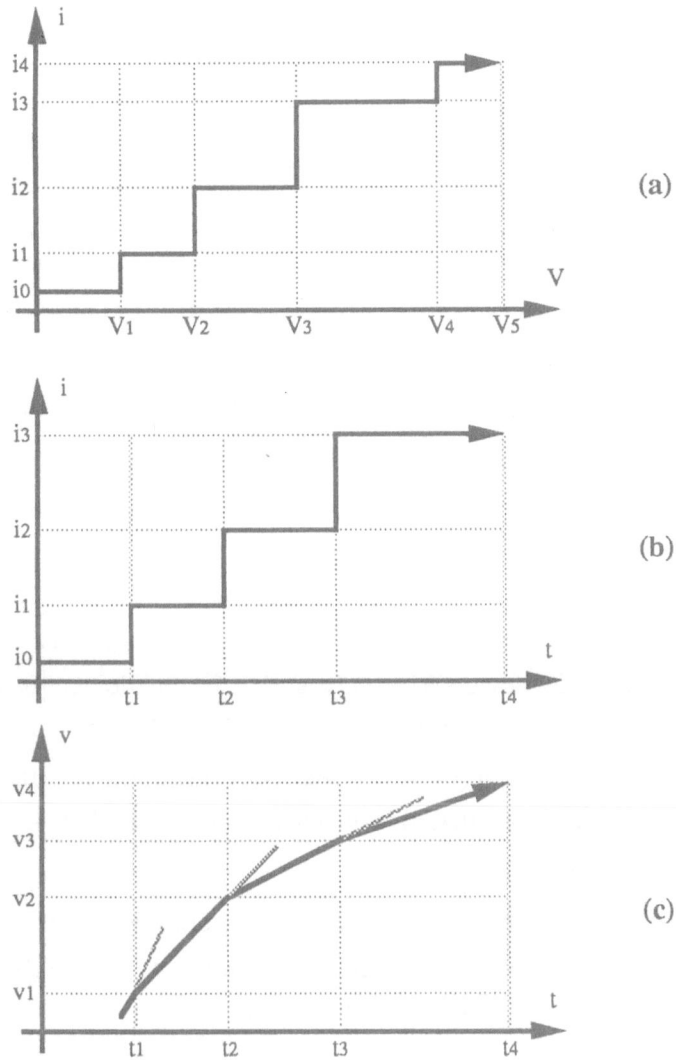

Figure 6.7: Table Models and Simulation Events in SPECS2

These steps are important in the event-driven approach of SPECS2 since an event occurs whenever a device reaches a "corner" of its step model, as shown in Figs. 6.7(b) and 6.7(c). Here, events occur at t_1, t_2 and t_3 since V_1, V_2 and V_3 are all boundaries of the device step model. For example, at t_2 there is a change in the current through the device from i_1 to i_2. As a result, the corresponding device is processed with the new current value and the next event is scheduled at the next corner in the table. The effect of this change on the rest of the circuit is taken into account via subsequent event scheduling and processing.

The SPECS approach is also prone to spurious oscillations, as are many other variable precision algorithms. The strategy used in SPECS2 to overcome this problem is to place the element into a pseudo-steady-state condition. This is done by picking a current for the device which is *in between* the currents in the table model whenever the direction of the current derivative (with respect to time) changes sign. The value is selected to place the device in steady state. If the device is truly in steady state, it will remain in this condition. On the other hand, if it is not, it will be forced out of it by the other elements in the circuit. Therefore, true circuit oscillation will not be suppressed but algorithmic oscillation will be prevented.

The iDSIM program uses macromodeling combined with waveform relaxation to perform switch-level timing simulation. The switch-level network is preprocessed to identify series-parallel connections of transistors to form composite transistors. A set of macromodel parameter tables is generated for each composite transistor based on the device gate voltage, device size, threshold voltage, and other factors that affect delay. The actual simulation is performed in two steps. First the circuit is analyzed using switch-level techniques to identify the transitions that will occur during the simulation. These transitions produce break points in the waveforms that, in turn, define the time intervals for detailed simulation. The second step is to perform the delay analysis to compute the transition times using the tables generated for the devices. If there are no feedback loops in the circuit, one pass of this algorithm is sufficient to produce the desired results. When feedback loops are present, a waveform relaxation approach is used with partial waveform convergence to compute the final results.

6.6. THE MIXED-MODE INTERFACE

A major issue in all mixed-mode simulators is the problem of interfacing of two or more simulation modes. This problem arises only when elements from different modes of simulation are connected at a common node. There are two possibile directions of signal conversion: one from the lower level of simulation (more detailed) to the higher level of simulation (less detailed), and a second going in the opposite direction. For example, logic simulation and electrical simulation require signal conversions from logic to electrical simulation and from electrical to logic simulation. Typically, it is easier to translate a signal from a lower level of simulation to a higher level since the conversion involves removing unnecessary details from the signal. Signal conversions from higher levels to lower levels are more difficult to perform. In fact, it is the conversion of signals from the logic domain to the electrical domain that is most troublesome and this is the key problem addressed below.

Early mixed-mode simulators used elements called logic-to-voltage (LTV) converters and logic-to-current (LTI) converters [NEW78] to perform signal mapping across the logic to the electrical interface. LTV converters were used to translate logic signals that were either 0 or 1 to an equivalent electrical voltage. A finite transition time was added for rising or falling logic waveforms to avoid convergence problems in the electrical algorithms. Because the input resistance of an ideal LTV is zero, this model was only adequate for driving high impedance loads, such as the gate node of an MOS transistor. The LTI converter was used at the interface whenever it was necessary to model the current-sourcing or current-sinking properties of a logic gate. An ideal LTI has an infinite input impedance and is, therefore, suited to driving low impedance loads such as the base of a bipolar transistor. These two converters are illustrated in Fig. 6.8.

The DIANA program [ARN78] introduced the concept of the Boolean-controlled switch (BCS) model where the state of a logic element was used to select one of two linear equivalent models. This was an important step in the modeling of the logic to electrical interface. The BCS model depicted in Fig. 6.9 selects the **R0–E0** model if the output is falling and the **R1–E1** model if the output is rising. The chosen model is presented to the electrical portion of the circuit and the node is then processed as part of electrical simulation. The values of the elements in the two Thevenin equivalent models can be adjusted to improve the accuracy, but the overall accuracy of this approach is limited. To understand the reason for this, the LTV, LTI and BCS converters are all shown in the same **R–V** plane in Fig. 6.10. Here, the inverter is assumed to be represented at the logic level and the pass transistor at the electrical level. The LTV model appears as a line at the zero resistance level while the LTI model appears as a line at the infinite resistance level. Clearly, these two models do not adequately represent the dc output characteristics of the inverter and so the accuracy during transient analysis is expected to be poor. On the other hand, the BCS model appears as two points and allows more precision in following the transfer curve. Here, the BCS mode has been chosen for the inverter output only. Unfortunately, this model only provides limited accuracy

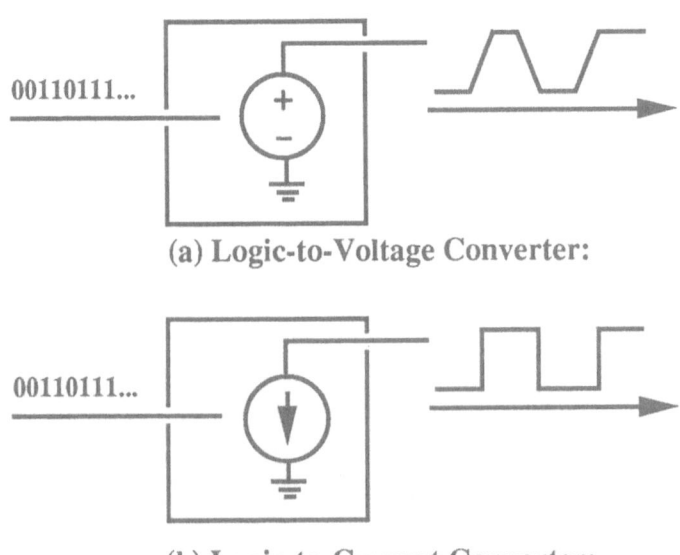

(a) Logic-to-Voltage Converter:

(b) Logic-to-Current Converter:

Figure 6.8: LTV and LTI Converters

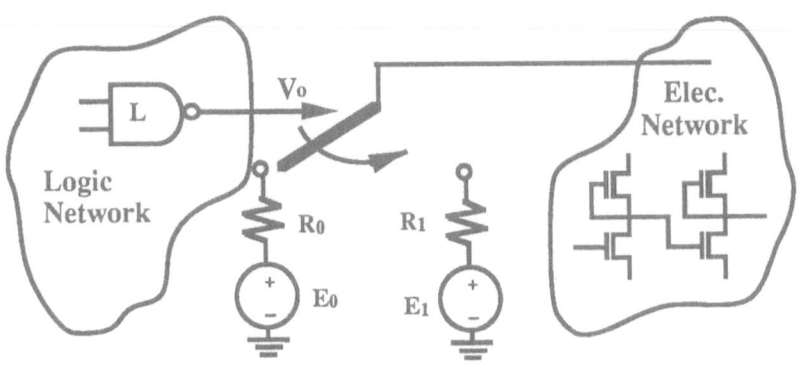

Figure 6.9: Boolean-Controlled Switch Model

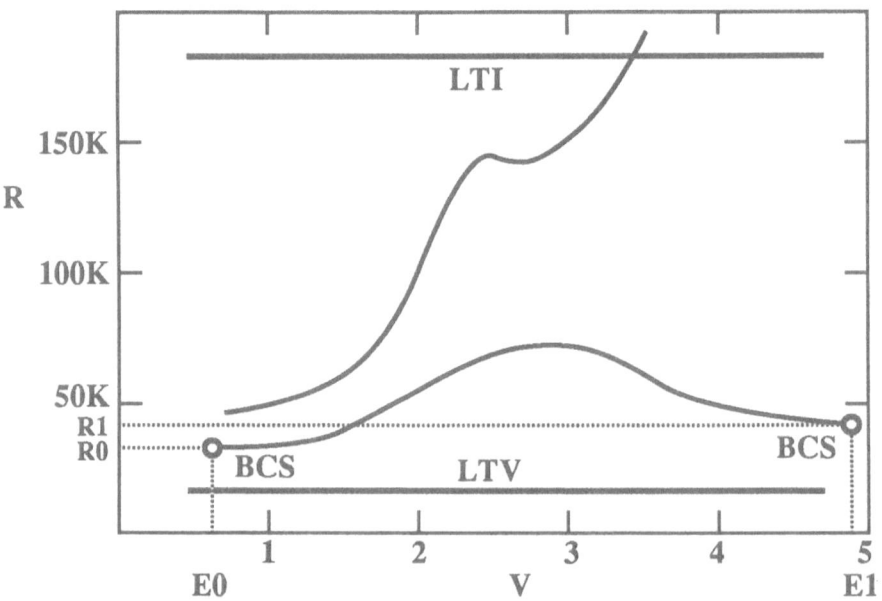

Figure 6.10: LTV, LTI and BCS in the **R–V** Plane

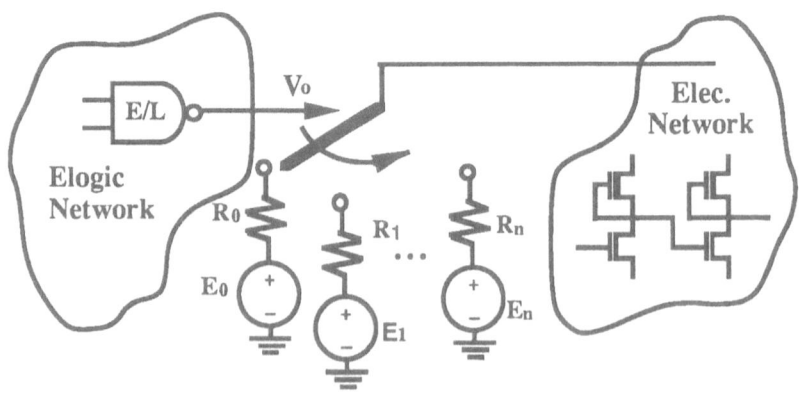

Figure 6.11: Voltage-Controlled Switch Model

during transient analysis since the features of the curve cannot be captured by two points alone.

One way to improve the accuracy of this approach is to allow more points along the trajectory of the dc transfer characteristic. This is the basic idea behind a generalization of the BCS model, called the voltage-controlled switch (VCS), as shown in Fig. 6.11. The use of this approach was first suggested in [KLE84] and is based on the Elogic modeling approach. Rather than choosing 1-out-of-2 models to represent the output, a choice of 1-out-of-n models is now available. The value of n depends on the number of voltage levels selected by the user. The greater the value of n, the better the accuracy. However, as shown earlier for the Elogic method, the CPU-time is proportional to the number of voltage states selected. Hence, a speed/accuracy tradeoff exists.

The generalized model is plotted on the $R-V$ plane in Fig. 6.12. A VCS model constructed from a five-state Elogic model is able to follow the trajectory of the curve quite closely; this greatly improves the accuracy during transient analysis. The transient performances of the BCS and VCS models are shown in Fig. 6.13 for the same example. When the inverter is represented using a BCS model, the results shown using the dotted lines are produced at node A. The five-state VCS model produces the results shown by the dashed lines. The actual results, derived using full electrical simulation, are also shown by a solid line. Clearly, both models adequately capture the details of the waveform at node A. However, the VCS based on the Elogic modeling approach provides better overall accuracy compared to the simple BCS. This interface modeling approach is now used in many mixed-mode simulators.

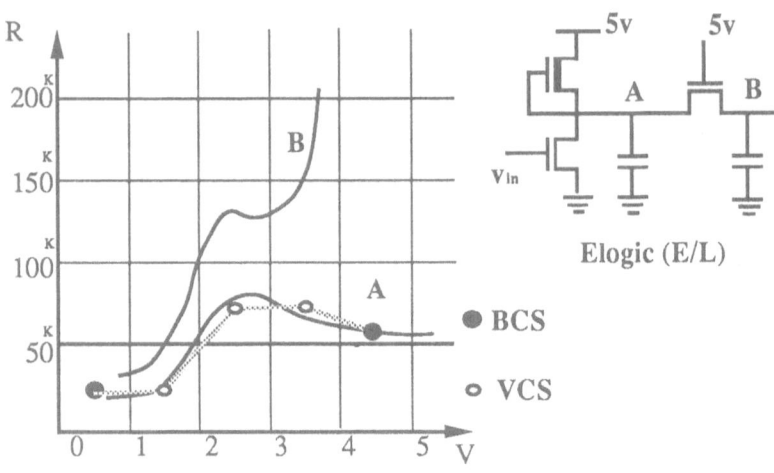

Figure 6.12: VCS Model on **R–V** Plane

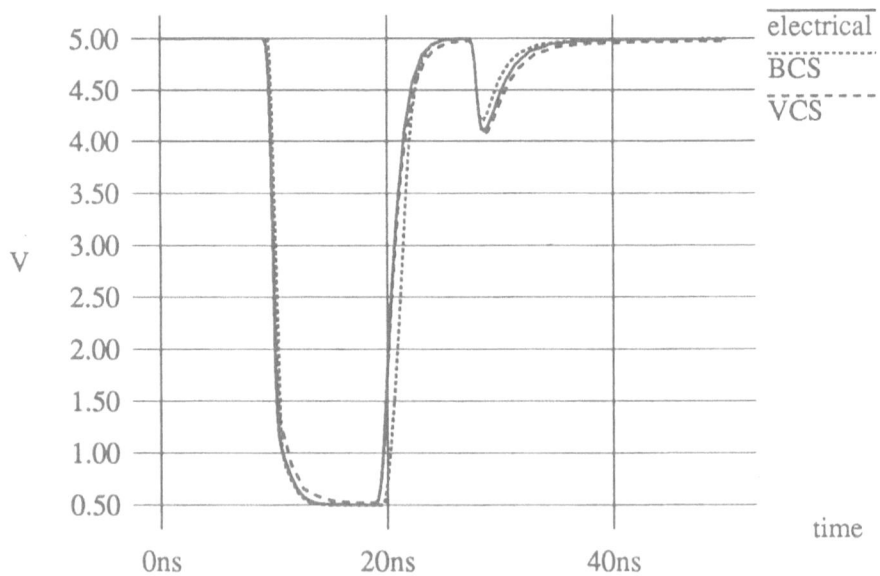

Figure 6.13: Transient Analysis Using BCS and VCS Models

CHAPTER 7

IMPLEMENTATION OF MIXED-MODE SIMULATION

This chapter describes the issues involved in implementing a mixed-mode simulator. The iSPLICE3 program [SAL89A] is used as a case study since it uses many of the algorithms described in the previous three chapters. The chapter begins with an overview of the architecture of iSPLICE3. Then, the issues associated with the implementation of an event scheduler are presented. Following this, the event scheduling issues during the transient analysis are described. Next, the techniques used to obtain the dc solution are provided. Finally, a number of simulation results using industrial examples are presented to indicate the performance of the program compared to SPICE2.

7.1. SIMULATOR ARCHITECTURE

To remain useful over its lifetime, a simulator must have the ability to expand and grow as the technology and simulation requirements evolve. To accomplish this, a simulator should be organized so that new algorithms and models can be easily added to the existing environment. Ideally, the addition of new algorithms or models should only involve a recompilation of the program to include the new routines. However, in practice, usually a few tables in a number of files must be modified to provide key pieces of information regarding the new models and algorithms.

In mixed-mode simulation, the use of event-driven, selective-trace in all modes of simulation is a unifying mechanism. To establish event-driven, selective-trace simulation, a time-queue and an event scheduler are required, and the notion of an *event* must be defined at each level of simulation. In iSPLICE3, each event data structure has associated *function*, *time* and *data* fields. When an event is processed, the *function* is performed on the *data* at the prescribed *time*. New events may be scheduled in the queue as part of the call to the function. Special simulation related tasks may also be scheduled in the time queue along with regular simulation events. This organization makes it relatively simple to add new simulation algorithms and functions since the structure of the events allows any type of function to be executed.

The basic simulation flow of iSPLICE3 is as follows:

```
main( )
{
        readin( );
        build_subcircuits();
        schedule ( setup_dc_analysis , t=0- );
        forall ( subcircuits S_i in the circuit)
            schedule ( S_i , t=0 ); /* for dc solution */
        schedule ( start_transient , t=0+ );
        /* MAIN SCHEDULER LOOP: */
        while (time queue is not empty) {
            event ← GetNextEvent();
            function ← event.simulation_Mode;
            time ← event.time;
            data ← event.simulation_Data;
            /* Perform task associated with event */
            function ( data, time );
        }
}
```

The circuit is first read in and divided into subcircuits during the *readin()* and *build_subcircuits()* phases. At the present time, the subcircuit types may be either LOGIC, ELOGIC, or ELECTRICAL, and are determined by the input circuit description provided by the user. iSPLICE3 determines the subcircuit type based on the devices connected to nodes in the circuit. For example, if a node has only ELOGIC devices connected to it, it will be labeled as an ELOGIC node. If it has only LOGIC devices connected to it, then it will be labeled as a LOGIC node. However, if there is at least one ELOGIC device controlling the node (i.e., the drain or source of a transistor), it will be labeled as an ELOGIC node. Similarly, if there is at least one ELECTRICAL device with a controlling node connected to it, it is labeled as an ELECTRICAL node. After the node assignments are completed, the ELECTRICAL nodes are further partitioned into subcircuits of tightly-coupled nodes as part of the standard ITA relaxation algorithm. Finally, the subcircuits and fanin and fanout tables are constructed using the node assignment information.

The next step is to schedule the *setup_dc_analysis()* event, and then schedule all the newly created subcircuits for evaluation at time t=0 as part of the dc solution. The last event to be scheduled before entering the processing loop is the *start_transient()* event, which is executed immediately after the dc solution is obtained. The program then enters the main loop where the scheduler sequences through the list of scheduled events. It remains in this loop until there are no events in the queue, at which time the program stops. The inner part of the loop involves obtaining the next event and then executing the function associated with the event. Examples of simulation functions scheduled in the time queue are *ELECTRICAL_event()*, *LOGIC_event()* and *ELOGIC_event()*.

The basic flow for a simulation event is shown below:

```
simulation_event( Sᵢ, t_N )
{
        get_input_voltages(Sᵢ );
        process_subcircuit(Sᵢ);
        if (Sᵢ is active)
            schedule ( Sᵢ , t_{N+1} );
        foreach ( node j in Sᵢ)
            foreach ( fanout subcircuit Sⱼ of node j )
                if (node j has crossed a critical threshold of Sⱼ)
                    schedule ( Sⱼ , t_N );
}
```

First the external voltages for the subcircuit are obtained. Then the subcircuit is processed using the appropriate analysis mode. If the voltages in the subcircuit have changed, the subcircuit is rescheduled for evaluation at a later time. Then the fanouts are scheduled at the current time if any important thresholds have been encountered. Other functions may also be scheduled in the same time queue for any special operations required during the simulation. Examples of such functions are *setup_dc_analysis()*, *setup_transient()*, *get_time_step()*, *backup_time()*, *wakeup_call()* and *process_pwl()*.

7.2. EVENT SCHEDULER DESIGN

In this section, a number of alternative strategies for the implementation of event schedulers are described. In designing a scheduler, a number of important issues relating to scheduler function and efficiency must be addressed. First, the event scheduler must have some notion of a time sequence and must be able to associate an event with a particular point in time. It may also be necessary to arrange events at a particular time point in some sorted order. Occasionally, the simulator will schedule an event and later decide that the event is no longer necessary. Hence, the scheduler must have the capability of canceling a pending event. Finally, the scheduling operations must be efficient, since they add to the simulation overhead. The event insertion/deletion operations must be relatively fast and the time sequencing through events should be efficient. Both of these requirements can be acheived by maintaining some uniformity in the event distributions in the scheduler, as will be seen. The scheduler overhead is usually insignificant for electrical analysis (since the events themselves are usually computationally intensive), but it may be a dominant factor in switch-level or higher levels of simulation where event processing operations are relatively simple. In general, the scheduling overhead should not consume more than 5-10% of the total simulation time. With these considerations in mind, the following implementations of event schedulers commonly found in mixed-mode simulators are presented.

7.2.1. Linear Linked-List Structure

To simplify the description initially, consider the situation where events may only be scheduled at integer multiples of a basic unit of time, Δt. The simplest structure for this type of scheduler is a *linear linked-list* of events in a time-sorted order as shown in Fig. 7.1. The list is usually referred to as a *time queue* and events are added to the queue by scanning through existing members of the list, starting at the present time (**PT**) pointer, and inserting them at appropriate points in the queue based on their respective *event times*. Events are processed in order starting with the event pointed to by **PT**. One problem with this approach is that the complexity of adding N events to the list is $O(N^2)$.

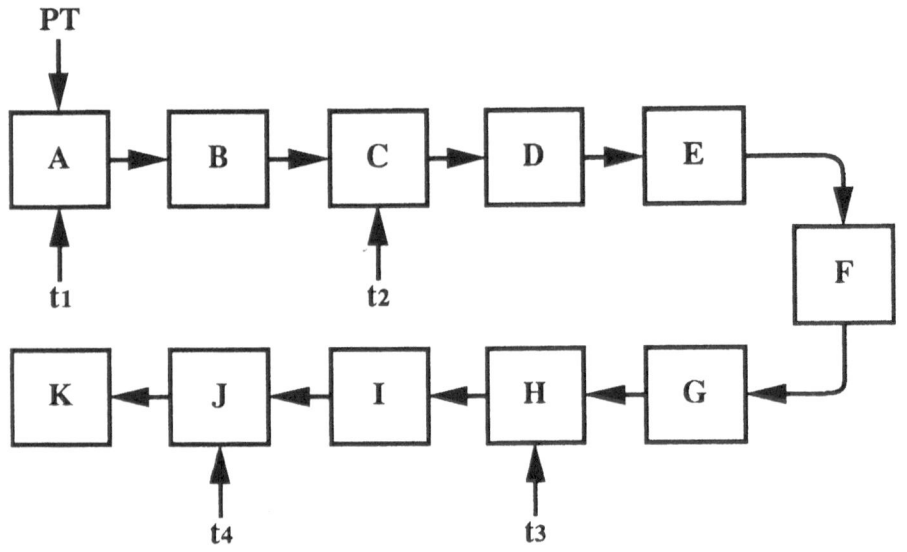

Figure 7.1: Linear Linked-List

A similar complexity exists when deleting events from the list.

7.2.2. Indexed List Methods

A somewhat more efficient organization, called a *linear indexed list*, is shown in Fig. 7.2 where the list contains headers that point to the true event list for each time point. These headers will be referred to as *time-point headers* (**H**) and the corresponding lists as *time-point event lists*. The time-point headers allow easy access to the time-point event lists where new events are to be added. As shown in the figure, the insertion process can be facilitated by including a *tail* pointer for each time-point event list. Therefore, adding an event is **O(H)** and removing an event is **O(H+N)**. Of course, the time-point headers must be scanned first to locate the correct list to insert the event. To avoid this, the headers can be organized as an array and indexed directly using the event time. This is referred to as an *array indexed list*. The method relies on the fact that the events are always scheduled at integer multiples of Δt. For example, an event at $PT+i\Delta t$ would be inserted into the time-point list which is i units from the current time pointer, PT. If a tail pointer is used to link events to the end of the list, the insertion time for each event is $O(1)$. Event cancellation involves indexing into the correct time-point event list and searching for the proper event to remove from the list. Hence, event cancellation is still $O(N)$ although the pre-multiplier of N is small.

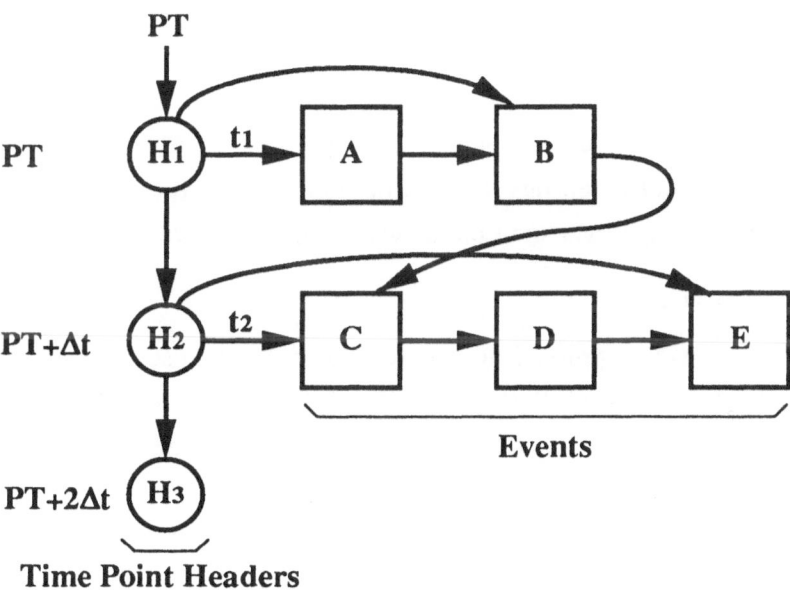

Figure 7.2: Linear Indexed List

The array indexed-list approach has the advantage that the correct header pointer can be accessed quickly. However, it cannot adapt to changing event distributions as easily as the linear indexed-list in Fig. 7.2. That is, the method is useful if the event distributions are uniform but may degenerate to the linear linked-list (Fig. 7.1) if the event distributions are nonuniform as shown in Fig. 7.3(a). If the event distribution becomes nonuniform, it is possible to adjust the Δt accordingly to produce a more uniform distribution, as shown in Fig. 7.3(b). This involves modifying all the header pointers in the array indexed-list to reflect the new Δt value and could be an expensive operation. In the linear indexed-list method, only the headers in the region of event congestion need to be modified to make the distribution more uniform. This method also accommodates varying Δt's from one header to the next. However, as stated earlier, event insertion is $O(N)$ in this approach due to the scanning process involved. The array-based approach requires scanning consecutive entries in the array only during the event processing phase and skips over time points where no events are scheduled.

7.2.3. Classical Time-Wheel

The classical approach to the scheduler design uses a *time-wheel* [BRE76] mechanism as illustrated in Fig. 7.4. This structure allows the indexed list to "wrap-around" so that the array of headers can be reused once the events associated with that entry have all been processed and the **PT** pointer has been incremented. For example, when the events at time **PT** have all been processed, the header at **PT** can be reused to represent the time **PT+MΔt**, assuming that the array has **M** elements. Using the MOD function, the **PT** pointer is always updated as follows:

$$PT = (PT + 1) \text{ MOD } M$$

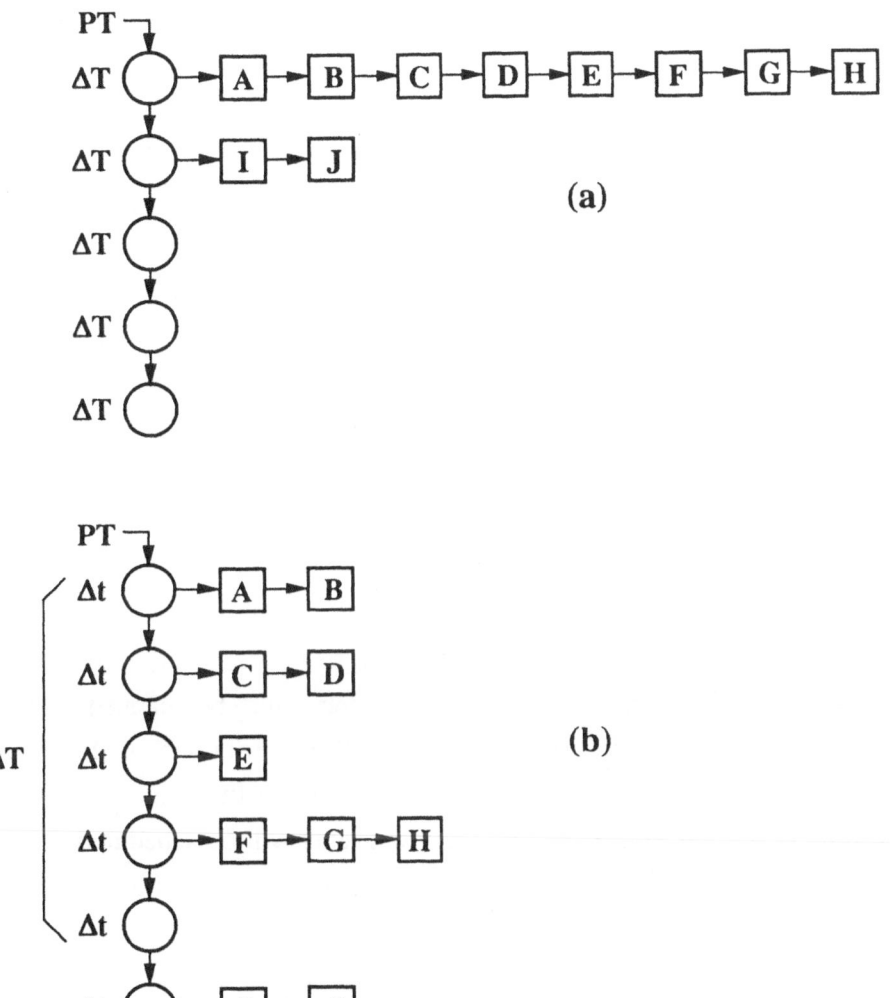

Figure 7.3: Nonuniform and Uniform Event Distributions

The obvious advantage of this approach is that a predetermined amount of memory can be allocated for the time queue *a priori*. However, dynamic memory allocation will still be required for events which occur at time points greater than t+MΔt units in the future. They may be organized in an *overflow* or *remote list*. At some point, these remote events must be brought into the time queue. Since the events in the remote list are usually somewhat more expensive to insert or remove than events in the time queue, it is not efficient to update the time queue with events from the remote list every time a time-point event list has been processed. However, as more and more time points are processed, the probability that new events will end up in the remote list increases, and this is undesirable. Therefore, it is better to move events from the remote list to the time queue periodically, i.e., after processing **k** time points in the time wheel.

Another source of inefficiency is due to the fact that many headers may not point to any events. These headers must be scanned anyway and this consumes additional CPU-time. The distribution of the events in the time queue, hence the sparsity, depends on the value of Δt. For example, if Δt is very small, only a few events will be scheduled at each time point, if any. On the other hand, if Δt is large, the events will densely populate the region of time near the current time pointer, **PT**. Both situations will reduce the scheduler performance. Hence, the number of time points processed before bringing in remote events (**k**), the size of the time wheel (**M**) and step size between adjacent entries in the time wheel (Δt), and indeed how efficiently the remote list is managed have an impact on the efficiency of this type of scheduler. Typical event distributions should be examined to select the appropriate values for these parameters for a given application.

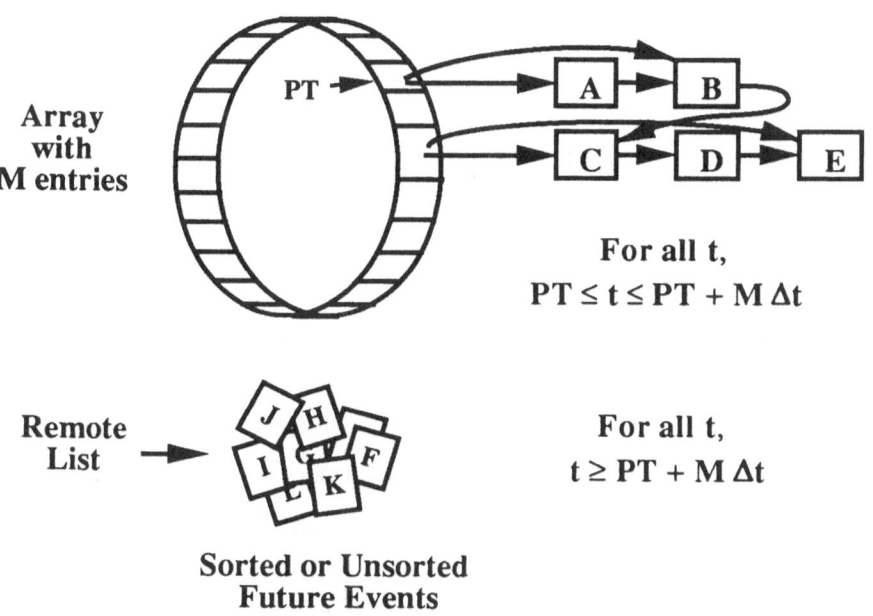

Array
with
M entries

For all t,
PT ≤ t ≤ PT + M Δt

Remote
List

For all t,
t ≥ PT + M Δt

Sorted or Unsorted
Future Events

Figure 7.4:   Classical Time-Wheel Mechanism

7.2.4. Managing Remote Lists

The remote list usually contains a small number of events if the proper parameter values are selected for the time queue. It usually contains events associated with external sources and these events are often sparse in time. The objective is to ensure that the ratio between events in the time queue and remote list does not exceed a certain threshold. If it is not anticipated that many events will be scheduled in the remote list, it may be organized as a simple linear linked list.

There are other situations where a more elaborate organization of the remote queue is required. In the case of electrical simulation, some components may take small time steps during a transition while others use very large time steps due to the latency. Here, a *secondary* time wheel would be useful. It can be managed in exactly the same way as the primary time wheel except that each interval is defined to be $k\Delta t$ units of time. After $k\Delta t$ units of time have been processed in the primary time queue, all the events in the next interval of the secondary queue can be moved to the primary queue. In general, it is possible to have a set of remote time wheels, each having an interval, Δt_i, which is equal to $k\Delta t_{i-1}$, where Δt_{i-1} is the interval used by the previous time-wheel. If a variable number of time wheels are used, another level of indexing would be useful in selecting the proper time-wheel.

The scheduler used in iSPLICE3 is similar to the classical time-wheel mechanism. However, rather than a single time-wheel, a pair of time queues with $M/2$ entries and a remote list are used. While events are being processed from the first queue, new events may be scheduled either in the remaining portion of the first queue, in the second queue or in the remote list. The remote list is maintained as a simple linear linked-list. When the end of the first queue is reached, the second queue becomes active and the first queue is adjusted to represent the next $(M/2)\Delta t$ units of time. Any appropriate remote events are moved from the remote list to the first queue. When the end of the second queue is reached, the first queue becomes the active queue again while the second queue is modified to represent the next $(M/2)\Delta t$ units of time. This scheme represents a compromise between bringing in new events after each time point list is processed and bringing in remote events only after all events in the queue have been processed.

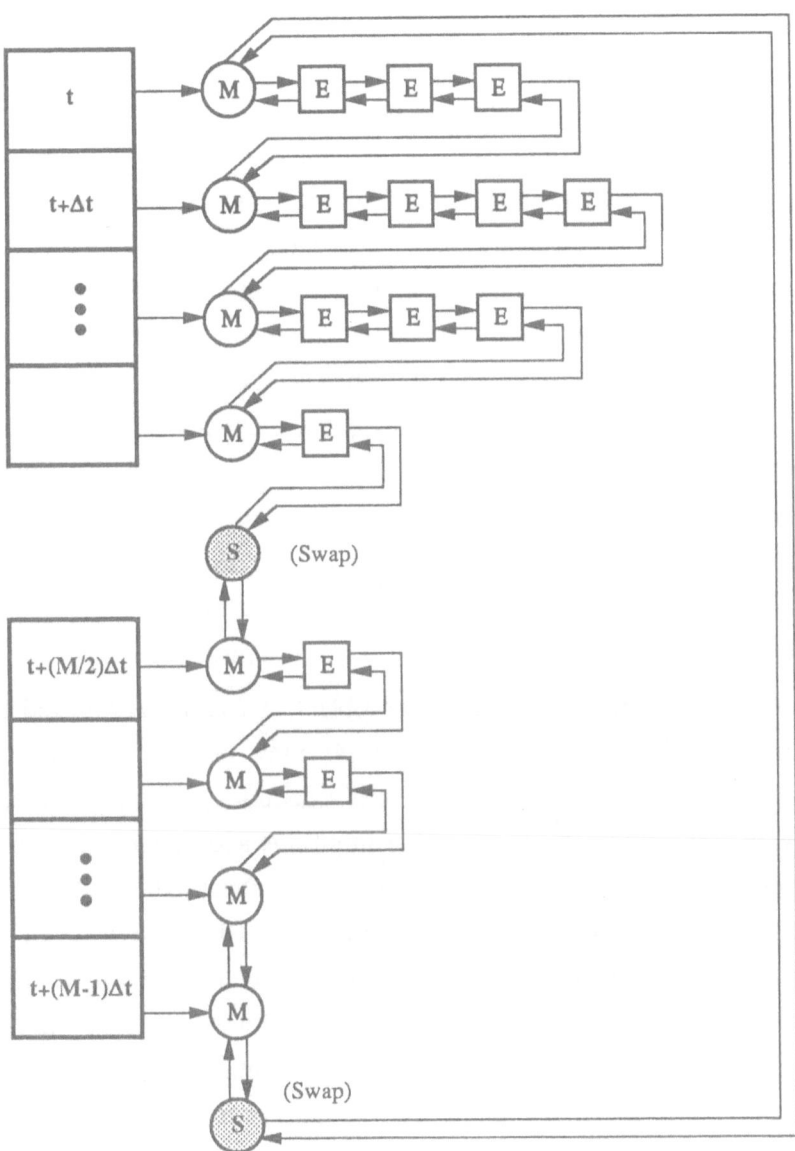

Figure 7.5: SPLICE Time Queue Data Structure

7.2.5. Other General Scheduling Issues

In the above description, it was assumed that the events are only scheduled at integer multiples of a basic unit of time, Δt. However, in electrical simulators, there may be a requirement to schedule events at virtually any point in time between 0 and the end of the simulation, T. In this situation, a few additional modifications must be made to the scheduler. The structures shown in Fig. 7.2 and Fig. 7.3 can still be used except that the Δt's now indicate the ranges of times of the events scheduled in a particular event list. That is, the events at time, t_{sched}, associated with the **jth** entry in the queue, actually lie in the range $t+j\Delta t \leq t_{sched} < t+(j+1)\Delta t$. Note that an event can no longer be scheduled at the end of a time point list, but instead, it must be inserted into the correct position according to its event time and the times of other events in the list. Clearly, the key to an efficient implementation for this case is to know *where* to start the search, as described below.

To suit the general scheduling problem, a technique referred to as the *cached-indexed list* method [KLE84] may be used. It uses the "locality of reference" property of event scheduling, meaning that existing events are usually clusted together in groups and new events are usually scheduled in the vicinity of previously scheduled events. In this approach, the headers are used as cache pointers which point to the *most recently* inserted event in a particular time interval. A cache pointer is used as a starting point for searches in the interval of the insertion, unless it happens to point to an event which is chronologically ahead of the event to be inserted. The cache pointer is then considered to be *invalid*. In that case, the headers are searched in reverse order until a *valid* cache pointer is found to begin the search. A valid cache pointer is defined as the first one which points to an event which precedes the event to be inserted. After adding the event, the original invalid header is updated to point to the new event. One of the benefits of this approach is that the method does not rely on the validity of the cache pointers. Therefore, if Δt is modified, the headers need not be adjusted. Instead, the same rules of scheduling are used, and as events are added or dropped from the list, the cache pointers will eventually be restored to some equilibrium condition. This approach was used in SPLICE2 [KLE84].

7.3. TRANSIENT ANALYSIS AND EVENT SCHEDULING

iSPLICE3 performs both dc analysis and time-domain, transient analysis of MOS and bipolar integrated circuits. Transient analysis is generally the most time-consuming and memory-intensive task in simulation but the mixed-mode techniques used in the iSPLICE3 program can reduce the simulation time significantly compared to that for SPICE2. iSPLICE3 has three simulation modes: circuit level simulation (ELECTRICAL) which uses iterated timing analysis, switch-level timing simulation (ELOGIC) and gate-level logic simulation (LOGIC). Each mode can be used independently or combined in a mixed-mode simulation. The details of each algorithm have been described in the previous chapters.

One issue that has been overlooked is that of event scheduling between different levels of simulation. A set of rules governing the scheduling policy from each simulation mode to other simulation modes must be defined. For example, one filtering operation that must be performed when processing ELECTRICAL subcircuits is to schedule their non-ELECTRICAL fanouts only when convergence occurs. This prevents non-ELECTRICAL fanouts from being processed unnecessarily with partial solutions during the iterations of ITA. However, other ELECTRICAL fanouts must be scheduled during the iterative process of

ITA. ELECTRICAL subcircuits schedule their LOGIC fanouts whenever a V_{IL} or V_{IH} threshold is encountered during an upward or downward transition, respectively. Similarly, ELECTRICAL subcircuits schedule their ELOGIC fanouts if they have encountered an ELOGIC state during the last transition. This is consistent with the scheduling used among ELOGIC subcircuits.

An ELOGIC subcircuit schedules its ELECTRICAL fanout subcircuits whenever it reaches a new ELOGIC voltage state. However, instead of actually scheduling an ELECTRICAL subcircuit at the current time, it simply ensures that the subcircuit is active by issuing a *wakeup_call()* event to any fanout ELECTRICAL subcircuits. If the fanout is not active, the *wakeup_call()* simply schedules it where the other ELECTRICAL subcircuits are scheduled. An ELOGIC subcircuit schedules its LOGIC fanouts if a V_{IL} or V_{IH} threshold has been encountered in its last voltage transition. LOGIC subcircuits schedule ELOGIC fanouts at each ELOGIC state along a transition of the logic waveform using *wakeup_call()*'s that are scheduled along transitions of logic waveforms. The same mechanism is used when LOGIC schedules ELECTRICAL subcircuits. Input source events follow similar rules as described above and are also dependent on the types of devices connected to them.

One additional complicating factor in intersimulation scheduling is due to *rollbacks* or step rejections. It may be necessary to occasionally cancel a pending event or reject a time-step and begin reprocessing at an earlier time. When this occurs, the scheduler must be backed up and the subcircuits rescheduled and reprocessed accordingly. The subcircuit which encountered the rejection is processed initially. If its new solution differs significantly from the previous one, its fanout subcircuits are scheduled. Otherwise, no scheduling operations are performed.

Similarly, the fanouts are processed and they compare their newly computed solutions with previous solutions and schedule their fanouts only if
the new solutions are different from their old solutions. Both ELOGIC
and LOGIC schedule events on fixed grid boundaries so that slight variations in the computed schedule times are not inferred as different solutions. The rollback strategy ensures that accurate solutions will be
obtained in an efficient manner.

7.4. DC ANALYSIS TECHNIQUES

iSPLICE3 provides a number of different techniques to obtain a dc
solution for a given circuit. For ELECTRICAL circuits, either the standard Newton method, source-stepping or gmin-stepping methods may be
invoked [QUA89]. For circuits that are represented using the ELECTRI-
CAL, LOGIC and ELOGIC levels, iSPLICE3 uses an iterative mixed-
mode dc solution scheme to initialize the node voltages, as follows:

```
dc_analysis()
{
    repeat {
        process_LOGIC_nodes(); /* using logic simulation */
        process_ELECTRICAL_and_ELOGIC_nodes();
        /* using Newton's method */
    } until (convergence)
    set_ELOGIC_nodes(); /* force to discrete values */
    repeat { /* correct any nodes affected by last operation */
        process_LOGIC_nodes();
        process_ELECTRICAL_nodes();
        /* leave out ELOGIC nodes */
    } until (convergence)
}
```
∎

The algorithm given above is performed at time 0 using event-driven techniques. First, the LOGIC nodes are processed using zero-delay logic simulation. Then the ELECTRICAL and ELOGIC nodes are processed using direct methods (i.e., the standard Newton method). Any nodes that are different from their previous solution act to schedule their fanout nodes at time 0. This process is repeated until convergence occurs. When the dc solution is obtained, the ELOGIC nodes are set to their nearest discrete values and the iterative loop is repeated once again to correct any values that may be affected by this operation. Unfortunately, the convergence of the dc solution is not guaranteed in all cases. In fact, if the LOGIC nodes do not have a dc solution, or if a proper initial guess is not specified for the ELECTRICAL and ELOGIC nodes, the iterative process may not converge at all!

While it is generally difficult to find a dc solution for LOGIC nodes that may oscillate when analyzed using zero-delay logic simulation, iSPLICE3 uses a new technique to improve the likelihood of convergence for ELECTRICAL and ELOGIC nodes in MOS digital circuits. This technique provides an initial guess that is usually close to the final solution, it ensures proper and reliable convergence and reduces the total number of Newton-Raphson iterations required. on MOS digital circuits described at the transistor level. First, the ELECTRICAL and ELOGIC portions of the circuit are solved using zero-delay, switch-level logic simulation [BRY80] to derive the initial conditions at each node. Then these logic values are converted to their corresponding voltage values. Next, the standard Newton method is applied to the same portion of the circuit, using the derived values as initial guesses. Since this technique provides an initial guess that is usually close to the final solution, it ensures proper and reliable convergence and reduces the number of overall iterations. This approach has been found to be 4-5 times faster than the standard approach on MOS digital circuits and successfully converges on circuits that fail to converge in SPICE2.

In the simple algorithm above, the processing of feedback paths deserves some special attention since all nodes are set to the uninitialized state as the first step of the switch-level analysis at time 0. iSPLICE3 processes the nodes from the inputs to the outputs, but if there are feedback paths in the network, some of the node values needed for the evaluation may be uninitialized, which presents a problem in determining the state of the output node. For these situations, iSPLICE3 guesses the values of initial unknowns whenever required as either logic 0 or logic 1, depending on the situation. If an incorrect guess is made, the feedback path will act to correct the situation in a subsequent processing step. This technique removes most of the uninitialized states at the output nodes, particularly in troublesome circuits such as flip-flops. However, some nodes may be assigned to the X state if the correct state can not be determined during switch-level simulation. These nodes are reset to 0 V before applying the Newton method since it places NMOS transistors in the cutoff region of operation rather than in some high-gain region.

As a simple example, consider the CMOS SR flip-flop circuit in Fig. 7.6. Assume that S=0 and R=1, and Q and $\overline{Q}$ are uninitialized. Then, if the upper NOR gate is processed by assuming that Q=1, a value of $\overline{Q}$=0 is produced. This value would be used to process the lower NOR gate and Q=0 is produced. Since this value is different from the original assumption, the first NOR gate is reprocessed to produce $\overline{Q}$=1, and the second reprocessed to produce Q=0. These are the correct solutions and so the processing would stop. Next, the values would be converted to their equivalent voltages and the Newton method would be invoked. A more interesing example is generated if S=0 and R=0 since

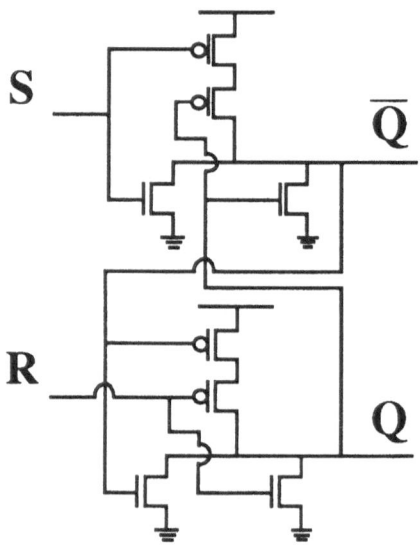

Figure 7.6: CMOS SR Flip-Flop Example

the previous outputs are held in the flip-flop for this case. Normally, a program like SPICE2 would produce values of Q=2.5V and $\overline{Q}$=2.5V (assuming a 5V supply voltage) as the dc solution, which is clearly incorrect. iSPLICE3 will produce either Q=0.0 and $\overline{Q}$=5.0 or Q=5.0 and $\overline{Q}$=0.0 and either case is an acceptable solution. Of course, the user can always override these values by initializing the flip-flops to any desired setting.

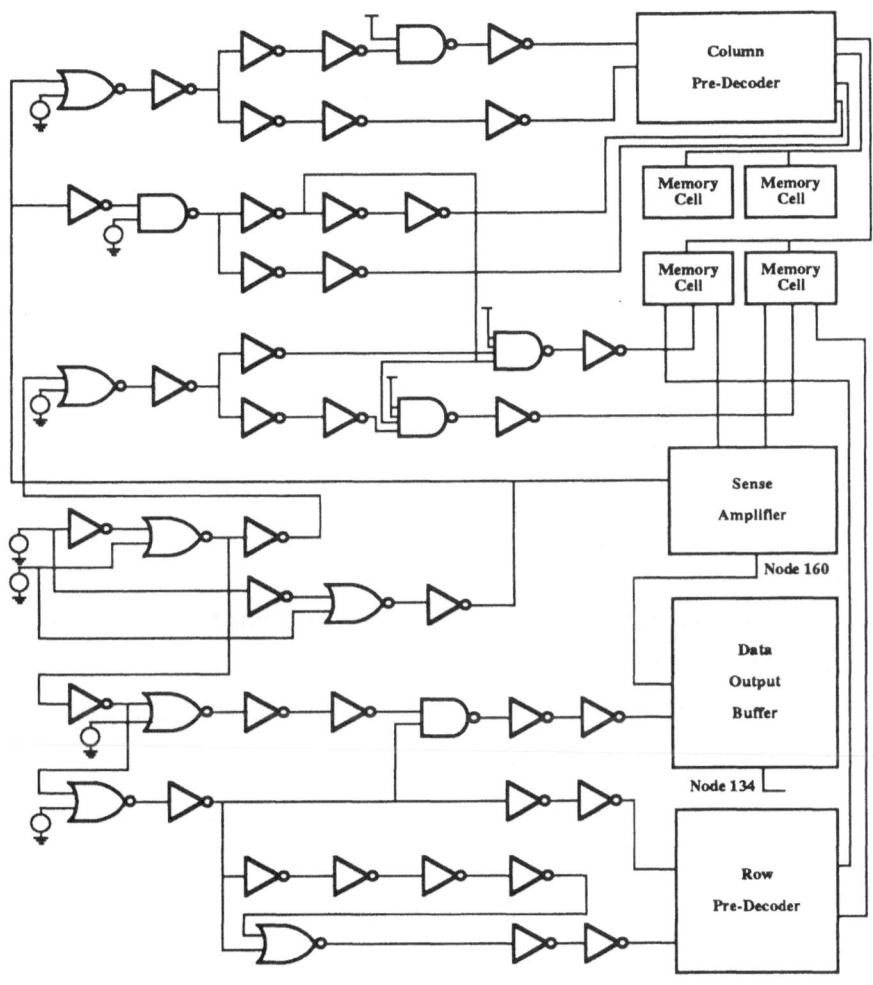

Figure 7.7: Block Diagram of a CMOS Static RAM Circuit

7.5. MIXED-MODE SIMULATION EXAMPLES

In this section, three industrial circuits that have been simulated using iSPLICE3 are presented and compared to SPICE2. The circuits are as follows: (1) a portion of a 64K CMOS static RAM, (2) a CMOS phase-locked loop used in a clock generation and synchronization application, and (3) a successive approximation A/D converter design in a BiCMOS technology. The three circuits are listed in Table 7.1 where the columns indicate the circuit name and simulation program used, the number of ELECTRICAL transistors, ELOGIC transistors, and LOGIC gates in the simulation, the total CPU-time and the speedup over SPICE2. Note that iSPLICE3 is listed twice for each circuit, once for ELECTRICAL simulation and once for mixed-mode simulation.

As the results indicate, iSPLICE3 is 30 to 40 times faster than SPICE2 for these circuits. Part of the speed improvement is provided by the ITA method which is 5 to 10 times faster than SPICE2. The remainder of the speedup is provided by mixed-mode simulation. In general, the speedup for a given circuit will depend on the number of transistors that are simulated at the ELECTRICAL level since this is the most expensive mode of simulation.

To illustrate the accuracy aspects of the program, consider the block diagram of a 64K CMOS Static RAM shown in Fig. 7.7. It is comprised of row decoders, column decoders, sense amplifiers, memory cells, etc., and is therefore ideal for mixed-mode simulation. The logic gates were carefully characterized for mixed-mode simulation using the corresponding transistor level circuits, and the appropriate parameter values for the gates were generated. Then, mixed-mode simulation was performed and two critical nodes were compared to the SPICE2 results: the output of the sense amplifier (node 160) and the data output bit (node 134). As Fig. 7.8 indicates, the results of the two simulations are almost indistinguishable. Note that the results for node 134 compare an electrical simulation in SPICE2 against Elogic simulation using 6-states in iSPLICE3, since Elogic was used for the output buffer circuitry. Therefore, the waveform features that are smaller than 1 volt are not present in the iSPLICE3 results. The results for the other nodes in this circuit, and the nodes in other circuits, have similar accuracy characteristics. Therefore, with proper attention to logic modeling and parameter extraction, the iSPLICE3 program can provide accurate simulation results with substantially shorter runtimes compared to SPICE2.

Program	Transistors		Gates	CPU-time	
	ELEC	ELOGIC	LOGIC	(sec.)	speedup
Static RAM					
SPICE2	277	0	0	3272	1
iSPLICE3	277	0	0	504	6.5
iSPLICE3	68	41	57	82.0	40
Phase-locked loop					
SPICE2	205	0	0	132000	1
iSPLICE3	205	0	0	20827	6.5
iSPLICE3	69	26	42	4117	32
A/D Converter					
SPICE2	1133	0	0	79200	1
iSPLICE3	1133	0	0	10830	7.5
iSPLICE3	243	294	175	2280	35

Table 7.1: Performance Comparisons Between iSPLICE3 and SPICE2
on a VAX 3500 Workstation

V

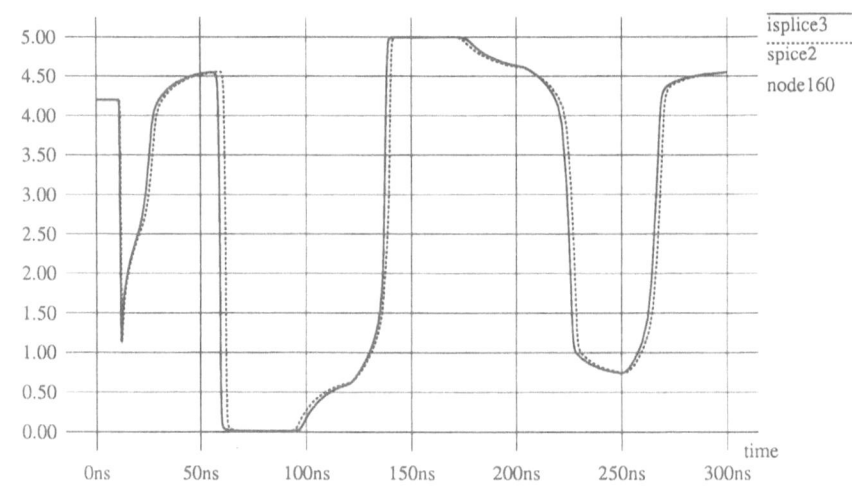

V

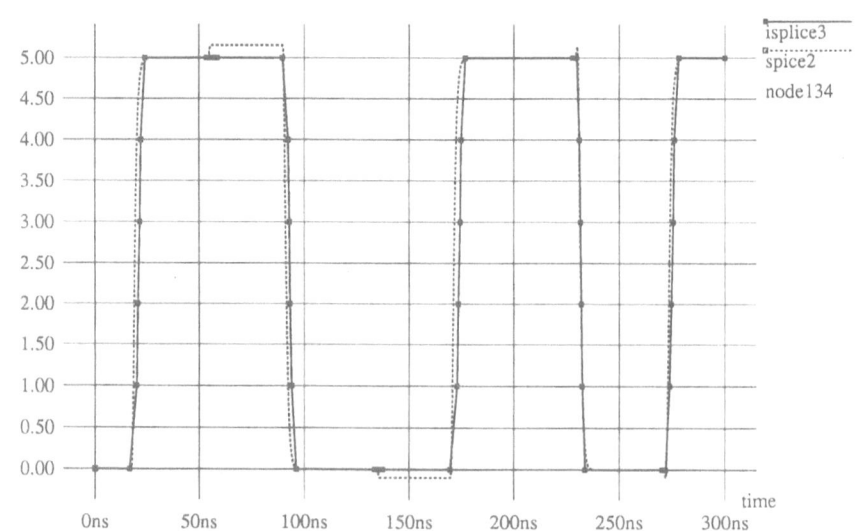

Figure 7.8: Accuracy Comparison Between iSPLCIE3 and SPICE2

CHAPTER 8

CONCLUSIONS AND FUTURE WORK

8.1. SUMMARY

A variety of techniques for mixed-mode simulation have been described in this book, focusing primarily on the combination of the gate-level, switch-level timing and electrical forms of simulation. Chapter 1 began with an overview of the different levels of simulation and provided the motivation for combining two or more levels into one simulator. Then, the basic issues in mixed-mode simulation were outlined and a brief survey of existing mixed-mode simulators was provided. In Chapter 2, the electrical simulation problem was formulated and the standard numerical techniques used to solve the problem were presented. The issues associated with the implementation of an efficient time-step control scheme were also described. In Chapter 3, two properties of waveforms, called *latency* and *multirate behavior*, were defined and used to motivate the need for new circuit simulation methods. Then, relaxation-based electrical simulation methods were introduced to exploit these waveform properties, and their theoretical aspects were described. Circuit partitioning methods to improve the convergence speed of relaxation methods were presented at the end of the chapter.

The electrical, gate-level and switch-level timing simulation algorithms were presented in Chapters 4, 5 and 6, respectively. These techniques make use of the event-driven, selective-trace paradigm which forms a common thread for all algorithms used in mixed-mode simulation. The main contribution of Chapter 4 was an event-driven circuit simulation algorithm to exploit latency. The evolution of logic states and logic delay models was presented in Chapter 5. The development of the Elogic simulation and modeling approach to resolve the signal mapping problems at the interface between electrical and logic elements was described in Chapter 6.

In Chapter 7, the implementation details of the iSPLICE3 mixed-mode simulator were presented. First, the overall architectural issues were described followed by a summary of the transient analysis techniques used and event scheduling policies enforced between the different levels of simulation. The issues associated with the implementation of event schedulers for mixed-mode simulators were also described in detail. A technique for the dc solution of mixed-level circuits was outlined. Finally, a number of mixed-mode simulation examples were provided.

8.2. AREAS OF FUTURE WORK

Although a substantial amount of work has been done in mixed-mode simulation, and much has been accomplished, there are still many promising areas of future work. In particular, automatic partitioning of transistor circuits, mixed-mode fault simulation, and analog multilevel simulation are topics of great interest in the research community. Each of these topics is outlined briefly in the sections to follow.

8.2.1. Automatic Partitioning

One problem not directly addressed in this book is that of defining the portions of a circuit to be represented at the various levels of abstraction. Normally, this task is the responsibility of the user since the circuit designer has the knowledge to perform the operation manually. However, one tool that would prove to be extremely useful in mixed-mode simulation is an automatic partitioner. Such a tool would be necessary when a transistor level description is extracted from a layout, or obtained from any other source, and a functional verification is desired in as short a time as possible. The partitioner would scan the circuit description and define the different levels of abstraction that would be used to simulate different portions of the circuit and then provide this information to the mixed-mode simulator.

Conceptually, this process takes a collection of components from a given level in the circuit description and replaces them with higher level primitives[1] to improve speed [RAO89, ACU89], or with lower level primitives to improve accuracy [OVE89]. The complete partitioning operation involves two phases: recognition and characterization. Groups of components that combine to form higher level primitives in the circuit must first be recognized using either a rule-based approach or a table lookup scheme. Then the parameters for the higher level model must be generated from the lower level description in the characterization phase to maintain simulation accuracy.

The following steps illustrate the process for the simple case of an MOS circuit. In the first phase, the circuit is divided into subcircuits of strongly-coupled (or channel-connected) components where transistors that are connected by their sources and drains are lumped into the same subcircuit. Next, each subcircuit is processed by examining the parallel and series connections of NMOS and PMOS devices to extract the logic function. Finally, the set of *electrical* transistors is replaced by an equivalent *logic* gate. In the second phase, the characterization process is performed. The parameters, such as V_{OH}, V_{OL}, t_r, and t_f, must be determined directly from the transistor level descriptions for every extracted gate using analytic equations, or possibly a few simulations.

---

[1] A primitive refers to a basic element that is known to the simulator, i.e., any element that is hard-coded into the program.

Any unrecognized structures are left to be simulated at the ELECTRI-CAL level during mixed-mode simulation.

There are a number of future directions to pursue in automatic partitioning. It would be useful to develop automatic methods to replace transistor level circuits with their gate equivalents for a variety of other technologies, including the bipolar and BiMOS technologies. Another important issue is to determine the highest level of simulation that can be used for a particular block of circuitry while still providing accurate results. This may require some short simulation runs of each block to determine its nature and simulation requirements. Along the same lines, it would be useful to determine the minimum number of Elogic states, and their corresponding voltage values, necessary to accurately simulate a given circuit. It would also be worthwhile to develop methods that can recognize higher level logic blocks such as flop-flops, registers, and adders from lower level primitives such as NANDs, NORs and inverters.

8.2.2. Fault Simulation

Fault simulation has been used in logic simulators to grade the quality of a set of input vectors in finding the potential faults in a circuit. However, the faults that can be modeled easily at the gate level are the simple *stuck-at* faults, such as the stuck-at-1 and stuck-at-0 faults [BRE76]. There are many other faults which are electrical in nature that are as important as, and in some cases more important than, the simple stuck-at faults in causing circuit malfunctions. For example, transient faults can occur from a number of sources including power supply transients, capacitive and inductive crosstalk and cosmic particle hits. Studies in this area have shown that approximately 80% of the computer system failures can be attributed to transient faults. Other permanent faults may arise due to shorts and opens between any two terminals of a bipolar or MOS transistor. Therefore, a study of electrically-oriented fault characteristics and their propagation is essential for the design of reliable digital systems.

Since a gate-level simulator cannot accommodate these faults, and a circuit simulator would be too expensive to use for fault simulation when the circuit is large, mixed-mode simulation can be used to address this problem. The electrically-oriented faults can be placed in any portion of a circuit by representing that portion at the electrical level while the rest of the circuit is represented at the gate level. This allows the simulator to run at the fastest possible speed while correctly modeling the true properties of the electrical fault. However, since thousands of faults must be simulated to obtain meaningful results, new techniques for concurrent mixed-mode fault simulation must be developed to make this approach feasible.

8.2.3. Analog Multilevel Simulation

The various levels of simulation discussed thus far are shown on the left side of Fig. 8.1. In defining this hierarchy of methods, there was a clear bias towards digital circuits, since they are usually large and tend to profit greatly from the higher levels of simulation. Analog circuit simulation was simply relegated to the lowest rung of the simulation ladder - electrical simulation. However, as shown on the right side of the figure, there is also a corresponding set of levels in the analog domain that has been overlooked until recently.

At the highest level is analog behavioral simulation where the individual blocks are described in terms of Laplace transforms or z-domain transfer functions and their interactions are described using signal flow diagrams that include summers, multipliers, etc. At the next level, circuit components, such as ideal opamps, switches, integrators, and

Digital   # Analog

Behavioral   **Behavioral**

RTL/ Gate   **Ideal Functional**

Timing   **Non-ideal Functional**

Electrical   **Electrical**

Figure 8.1: Levels of Simulation

comparators are used. This corresponds to the RTL level in digital circuits. Although the models are idealized, this level allows the designer
to validate a proposed architecture for the design. The next level is
similar to the previous level except that the first- and second-order
details are included in the models. For example, finite bandwidth, finite
gain, and input and output resistances would be included in the opamps,
and switch capacitances and resistances included in the MOS switches
when simulating switched-capacitor filter circuits. This corresponds to
gate and switch-level that includes timing information and other nonlinear effects. Finally, at the most detailed level, electrical simulation is

available, which corresponds directly to the same level on the digital side of the figure.

An important goal in analog simulation is to develop a multilevel simulation environment that incorporates all of these different levels of simulations and allows both time-domain and frequency-domain analyses. This would permit the designer to represent different portions of the design at any desired level in either the time-domain or the frequency-domain. The simulator would be responsible for transforming the representations from one domain to the other and resolving the issue of different levels of the hierarchy in one circuit description. Great progress has been made in this area recently [CHA89, SWI89] but much work remains, especially in the nonideal functional level of simulation to include nonlinearities and noise models. New macromodeling approaches must be developed that include these effects in standard analog blocks for use in a wide range of applications. Finally, the analog simulation levels should be combined with the digital simulation levels so that both of the hierarchies in Fig. 8.1 can be mixed and matched easily within a single simulation environment.

8.3. CONCLUSIONS

Mixed-mode simulation is now a well-accepted form of simulation in industry for large circuits containing both analog and digital components, and for circuits described at multiple levels of abstraction. A wide variety of simulators have been developed, in both industry and academia, and many are in production use today. As described in this book, the key contribution of mixed-mode simulation is that it offers the designer the ability to intelligently trade off simulation precision for simulator performance within the scope of a single simulator thereby permitting the designer to choose detailed simulation where accuracy is essential and higher forms of simulation where less accuracy can be tolerated. A second important theme of the book is that mixed-mode simulation provides a uniform environment for designers to develop ideas from initial concepts to the final circuit schematics and accommodates both top-down and bottom-up design styles, or any form in between. In addition, designers often mix different levels of abstraction in a single schematic diagram to convey the important aspects of a circuit design. These different representations can be captured easily in a mixed-mode simulation environment and later used to verify the circuit operation and performance. Finally, mixed-mode simulators are flexible and extensible and provide high performance in circuit verification. These features combine to make mixed-mode simulation one of the most powerful tools for VLSI design and simulation.

REFERENCES

[ACU89] E. Acuna, J. Dervenis, A. Pagones, R. Saleh, "iSPLICE3: A New Simulator for Mixed Analog/Digital Circuits", Custom Integrated Circuits Conference Digest of paper, May 1989, pp. 13.1.1-13.1.4.

[AGR80] V.D. Agrawal, A.K. Bose, P. Kozak, H.N. Nham, "A Mixed-Mode Simulator", Proc. of 17th Design Automation Conference, pp. 618-625, June 1980.

[ARN78] G. Arnout, H. DeMan, "The Use of Thresholding Functions and Boolean-Controlled Elements for Macromodelling of LSI Circuits", IEEᴸ J. of Solid-State Circuits, SC-13, June 1978, pp. 326-332.

[BEA86] J. M. Beardslee, "Implementation of a Logic Simulator and Mixed-Level Simulation for SAMSON2", M. S. Report, Carnegie-Mellon University, Report No. CMUCAD-86-12, May 1986.

[BRA72] R.K. Brayton, F.G. Gustavson, G.D. Hachtel, "A New Efficient Algorithm for Solving Differential-Algebraic Systems Using Implicit Backward-Differentiation Formulas", Proceedings of the IEEE, Vol. 60, No. 1, pp. 98-108, Jan. 1972.

[BRE72] M.A. Breuer, "A Note on Three-Valued Logic Simulation," IEEE Trans. on Computers, April 1972, pp. 399-402.

[BRE75] M.A. Breuer, Ed., **Digital System Design Automation: Languages, Simulation and Data Base**, Computer Science Press, 1975.

[BRE76] M.A. Breuer and A. D. Friedman, **Diagnosis and Reliable Design of Digital Systems**, Computer Science Press, 1976.

[BRY80] R. E. Bryant, "An Algorithm for MOS Logic Simulation", LAMBDA, 4th Quarter 1980, pp. 46-53.

[BRY84] R. E. Bryant, "A Switch-Level Model and Simulator for MOS Digital Systems", IEEE Trans. on Computers, Vol. c-33, no. 2 , Feb. 1894, pp. 160-177.

[BRY87] R. E. Bryant, "A Survey of Switch-Level Algorithms", IEEE Design and Test of Computers, vol. 4, no. 4, Aug. 1987, pp. 26-40.

[BUR83] J. L. Burns, A. R. Newton, D. O. Pederson, "Active Device Table Lookup Models for Circuit Simulation", International Symposium on Circuits and Systems, Newport Beach, CA, May 1983.

[CAR84] C.H. Carlin, A. Vachoux, "On Partitioning for Waveform Relaxation Time-Domain Analysis of VLSI Circuits", International Symposium on Circuits and Systems, Montreal, Canada, May 1984.

[CHU75] L. Chua, P. Lin, **Computer-Aided Analysis of Electronic Circuits: Algorithms and Computational Techniques**, Prentice-Hall, 1975.

[CHA75] B.R. Chawla, H.K. Gummel, and P. Kozak, "MOTIS - An MOS Timing Simulator," IEEE Trans. Circ. and Sys., Vol. 22, pp. 901-909, 1975.

[CHA87] H.P. Chang, J. A. Abraham, "The Complexity of Accurate Logic Simulation", Int. Conf. on Computer-Aided Design, Santa Clara, CA., Nov. 1987, pp. 404-407.

[CHA89] T. Chanak, R. Chadha, , K. Singhal, "Switched-Capacitor Simulation for Full-Chip Verification," Proc. of the Custom Integrated Circuits Conference, San Diego, CA., May 1989.

[CHE84A] C. F. Chen, P. Subramaniam, "The Second Generation MOTIS Timing Simulator-- An Efficient and Accurate Approach for General MOS Circuits" International Symposium on Circuits and Systems, Montreal, Canada, May 1984.

[CHE84B] C. F. Chen, C-Y Lo, H.N. Nham, P. Subramaniam, "The Second Generation MOTIS Mixed-Mode Simulator", Proc. of 21nd Design Automation Conference, pp. 10-17, June 1984.

[COR88] T. Corman, "Using VIEWSIM/AD To Simulate Mixed Analog and Digital Systems", Electro/88, Session 43 Record, Boston, MA. May 1988.

[DEG84] A. DeGeus, "SPECS: Simulation Program for Electronic Circuits and Systems," Proc. IEEE Int. Symp. on Circ. and Sys., pp. 534-537, May 1984.

[DEM81A] H. DeMan, G. Arnout, P. Reynaert, "Mixed-mode Circuit Simulation Techniques and Their Implementation in DIANA", NATO Advanced Study Series, in **Computer-Aids for VLSI Circuit**, Sijthoff & Noordhoff International Publishers, The Hague, pp. 113-174, 1981.

[DEM80] G. De Micheli, " New Algorithms for the Timing Analysis of MOS Circuits" Master Report, University of California, Berkeley, 1980.

[DEM81B] G. De Micheli, A. Sangiovanni-Vincentelli, "Numerical Properties of Algorithms for the Timing Analysis of MOS VLSI Circuits", University of California, Berkeley, ERL Memo. UCB/ERL M81/25, May 1981.

[DEM83] G. De Micheli, A.R. Newton, A. Sangiovanni-Vincentelli, "Symmetric Displacement Algorithms for the Timing Analysis for VLSI MOS Circuits", IEEE Trans. on Computer-Aided Design, Vol CAD-2, No. 3, pp. 167-180, July 1983.

[DES69] C.A. Desoer, E.S. Kuh, *Basic Circuit Theory*, McGraw-Hill, 1969.

[DUM86] D. Dumlugol, "Segmented Waveform Relaxation Algorithms for Mixed-Mode Simulation of Digital MOS VLSI Circuits", Ph.D. Dissertation, Katholieke Universiteit Leuven, Oct. 1986.

[EIC65] E.B. Eichelberger, "Hazard Detection in Combinational and Sequential Switching Circuits", IBM J. Res. and Develop., Vol. 9, March 1965, pp. 90-99.

[FAN77] S. P. Fan, M. Y. Hsueh, A. R. Newton and D. O. Pederson, "MOTIS-C A new circuit simulator for MOS LSI circuits," International Symposium on Circuits and Systems, April 1977.

[GEA71] C. W. Gear, **Numerical Initial Value Problems in Ordinary**

Differential Equations, Prentice-Hall, Englewood Cliffs, N.J., 1971.

[GEA80] C. W. Gear, "Automatic Multirate Methods for Ordinary Differential Equation", Information Processing 80, International Federation of Information Processing, 1980.

[GRE88] S. Greenberg, J. Grodstein, K. Sakallah, "Mixed Analog-Digital Simulation", Electro/88, Session 43 Record, Boston, MA. May 1988.

[GRO87] J. J. Grodstein, T.M. Carter, "SISYPHUS - An Environment for Simulation", Proc. Int. Conf. on CAD, Santa Clara, CA., Nov. 1987, pp. 400-403.

[GYU85]. R.S. Gyurscik, "A MOS Transistor Model-Evaluation Attached Processor For Circuit Simulation", Proc. IEEE Int. Conf. on Computer-Aided Design, Santa Clara, CA., Nov. 1985.

[HAC71] G.D. Hachtel, R.K. Brayton, F.G. Gustavson, "The Sparse Tableau Approach to Network Analysis and Design", IEEE Trans. on Circ. Theory, Vol. CT-18, pp. 101-113, Jan. 1971.

[HEN85] B. Hennion, P. Senn, "ELDO: A New Third Generation Circuit Simulator Using the One-step Relaxation Method" International Symposium on Circuits and Systems, Kyoto, Japan, June 1985.

[HIL80] D. Hill, "Multilevel Simulator for Computer-Aided Design", Ph.D. dissertation, Dept. of Elec. Eng., Stanford University, 1980.

[HO75] C.W. Ho, A.E. Ruehli, P.A. Brennan, "The Modified Nodal Approach to Network Analysis", IEEE Trans. on Circ. and Sys., Vol. CAS-22, pp. 504-509, June 1975.

[HSI85] H.Y. Hsieh, A.E. Ruehli, P. Ledak, "Progress on Toggle: A Waveform Relaxation VLSI-MOSFET CAD Program" International Symposium on Circuits and Systems, Kyoto, Japan, June 1985.

[HUA83] T. Huang, "Analysis of a Method for the Timing Simulation of Large-Scale MOS Circuits Containing Floating Capacitors" Master Report, University of California, Berkeley 1983.

[INF84] B. Infante, A. Sanders, E. Lock, "Hierarchical Modeling in a Multi-level Simulator", International Conference on Computer-Aided Design, pp. 39-41, Santa Clara, CA. 1984.

[INS84] A. Insinga, "Behavioral Modeling in a Structural Logic Simulator", International Conference on Computer-Aided Design, pp. 42-44, Santa Clara, CA. 1984.

[KIM84] Y. Kim, J.E. Kleckner, R.A. Saleh, A.R. Newton, "Electrical-Logic Simulator", International Conference on Computer-Aided Design, Santa Clara, CA., pp. 7-10, Nov. 1984.

[KLE83] J.E. Kleckner, R.A. Saleh, A.R. Newton, "Electrical Consistency in Schematic Simulation", International Conference on Circuits and Computers, NY, October 1983.

[KLE84] J. E. Kleckner, "Advanced Mixed-Mode Simulation

Techniques", Ph.D. dissertation, University of California, Berkeley, May 1984.

[KUN86] K. S. Kundert, "Sparse Matrix Techniques and their Application to Circuit Simulation", **Circuit Analysis, Simulation and Design**, A.E. Ruehli, ed., North-Holland Pub. Co., 1986.

[LEE88] E. S. Lee, T-F Fang, "A Mixed-Mode Analgo-Digital Simulation Methodology for Full Custom Design", Custom Integrated Circuits Conference, pp. 3.5.1-3.5.4, May 1988.

[LEL82] E. Lelarasmee, A. E. Ruehli, A. L. Sangiovanni-Vincentelli, "The Waveform Relaxation Method for Time-Domain Analysis of Large Scale Integrated Circuits," IEEE Trans. on CAD of IC and Sys., Vol. 1, n. 3, pp.131-145, July 1982.

[MAR85] G. Marong and A. Sangiovanni-Vincentelli, "Waveform Relaxation and Dynamic Partitioning for Transient Simulation of Large Scale Bipolar Circuits" International Conference on Computer-Aided Design, Santa Clara, CA, Nov. 1985.

[MCC88] W.J. McCalla, **Computer-Aided Circuit Simulation Techniques**, Kluwer Academic Publishers, Boston, MA. 1988.

[NAG75] L.W. Nagel, "SPICE2: A Computer Program to Simulate Semiconductor Circuits," Electronics Research Laboratory Rep. No. ERL-M520, University of California, Berkeley, May 1975.

[NAG80] L.W. Nagel, "ADVICE for Circuit Simulation," International

Symposium on Circuits and Systems, May 1980.

[NEW78A] A. R. Newton, D. O. Pederson, "Analysis Time, Accuracy and Memory Tradeoffs in SPICE2", 12th Asilomar Conference on Circuits, Systems and Computers, Asilomar CA, November 1978.

[NEW78B] A.R. Newton, "The Simulation of Large-Scale Integrated Circuits", Ph.D. dissertation, University of California, Berkeley, ERL Memo. ERL-M78/52, July 1978.

[NEW79] A. R. Newton, "The Analysis of Floating Capacitors for Timing Simulation," Proc. 13th Asilomar Conf. on Circ., Sys. and Comp., Asilomar CA, November 1979.

[NEW81] A.R. Newton, "Timing, Logic and Mixed-Mode Simulation for Large MOS Integrated Circuits", in **Computer-Aids for VLSI Circuits**, Sijthoff & Noordhoff International Publishers, The Hague, pp. 175-239, 1981.

[NEW83] A.R. Newton, A. Sangiovanni-Vincentelli, "Relaxation-based Circuit Simulation", IEEE Trans. on Elec. Dev., Vol. ED-30, No. 9, pp. 1184-1207, Sept. 1983.

[ODR86] P. Odryna, K. Nazareth, C. Christensen, "A Workstation-based Mixed-Mode Circuit Simulator", Proc. of the 23rd Design Automation Conference, pp. 186-191, June 1986.

[ORT70] J. M. Ortega and W.C Rheinbolt, **Iterative Solution of Nonlinear Equations in Several Variables**, Academic Press, 1970.

[OVE88] D. Overhauser, I. Hajj, "A Tabular Macromodelling Approach to Fast Timing Simulation Including Parasitics," International Conference on Computer-Aided Design, Santa Clara, CA., pp. 70-73, 1988.

[OVE89] D. Overhauser, "Fast Timing Simulation of MOS VLSI Circuits", Ph.D. Dissertation, University of Illinois, Aug. 1989.

[PEN81] P. Penfield, J. Rubenstein, "Signal Delays in RC Tree Networks," Proc. of 18th Design Automation Conference, pp.613-617, June 1981.

[QUA89] T. Quarles, "Analysis of Performance and Convergence Issues for Circuit Simulation," Ph.D. Dissertation, UCB/ERL M89/42, University of California, Berkeley, April 1989. Berkeley, CA. 1989.

[RAO85] V. Rao, "Switch-level Timing Simulation of MOS VLSI Circuits", Ph.D. dissertation, University of Illinois, UILU-ENG-85-2207, R-1032, Jan. 1985.

[RAO89] V. Rao, D. Overhauser, I. Hajj, T. Trick, **Switch-level Timing Simulation of MOS VLSI Circuits**, Kluwer Academic Publishers, Boston, MA., 1989.

[RAB79] N.B.G. Rabbat, A. Sangiovanni-Vincentelli and H.Y Hsieh, "A Multilevel Newton Algorithm with Macromodelling and Latency for the Analysis of Large-Scale Nonlinear Circuits in the Time Domain", IEEE Trans. on Circ. and Sys., Vol. CAS-26, pp.733-741, Sep. 1979.

[SAL83] R. A. Saleh, J. E. Kleckner and A. R. Newton, "Iterated

Timing Analysis and SPLICE1", International Conference on Computer-Aided Design, Santa Clara, CA., 1983.

[SAL84] R. Saleh, "Iterated Timing Analysis and SPLICE1", Master Report, University of California, Berkeley, 1984.

[SAL89A] R. Saleh, "iSPLICE3 User's Guide", University of Illinois, 1989.

[SAL89B] R. Saleh, A. R. Newton, "The Exploitation of Latency and Multirate Behavior using Nonlinear Relaxation for Circuit Simulation," IEEE Trans. on Computer-Aided Design of Circ. and Sys., Dec. 1989.

[SAK80] K. Sakallah and S.W. Director, "An Activity-Directed Circuit Simulation Algorithm," International Conference on Circuits and Computers, October 1980.

[SAK81] K.A. Sakallah, "Mixed Simulation of Electronic Integrated Circuits", Ph.D. dissertation, Carnegie-Mellon University, DRC-02-07-81, Nov. 1981.

[SAK85] K. Sakallah, "Polynomial Terminal Equivalent Circuits as Dormant Models in Event Driven Circuit Simulation", International Conference on Computer-Aided Design, Santa Clara, CA, 1985.

[SAN77] A. Sangiovanni-Vincentelli, L.K. Chen and L.O. Chua, "A New Tearing Approach-Node Tearing Nodal Analysis", International Symposium on Circuits and Systems, 1977.

[SZY75] S.A.Szygenda and E.W.Thompson, "Digital Logic Simulation in a Time-Based, Table-Driven Environment. Part 1. Design Verification," IEEE Computer Magazine, March 1975, pp.24-36.

[SWI89] SWITCAP-II Users Guide, Columbia University, 1989.

[TAH87] H. Tahawy, G. Mazare, B. Hennion, P. Senn, "A New Implementation Technique for the Simulation of Mixed (Digital-Analog) VLSI Circuits," International Conference on Computer-Aided Design, Santa Clara, CA. Nov. 1987, pp. 396-399.

[TER83] C. Terman, "RSIM - A Logic-Level Timing Simulator", Int. Conf. on Comp. Design, Port Chester, NY, 1983.

[TSA85] D. Tsao, C-F Chen, "A Fast Timing Simulation for Digital MOS Circuits", International Conference on Computer-Aided Design, Santa Clara, CA. Nov. 1985, pp. 185-187.

[UYE88] J.P. Uyemura, **Fundamentals of MOS Digital Integrated Circuits**, Addison-Wesley Pub., 1988.

[VAR62] R. S. Varga, **Matrix Iterative Analysis**, Prentice-Hall, 1962.

[VID86] L. Vidigal, S. Nassif, S. Director, "CINNAMON: Coupled Integration and Nodal Analysis of MOS Networks," 23rd Design Automation Conference, pp. 179-185, June 1986.

[VIS86] C. Visweswariah, "SPECS2: A Timing Simulator", M.S. Report, Carnegie-Mellon University, Report No. CMUCAD-86-24,

October 1986.

[WAR78] D.E. Ward and R.W. Dutton, "A Charge-Oriented Model for MOS Transient Capacitances", IEEE J. Solid-state Circuits, vol. SC-13, Oct. 1978.

[WEE73] W. T. Weeks, A. J. Jimenez, G. W. Mahoney, D. Mehta, H. Qassemzadeh, and T. R. Scott, "Algorithms for ASTAP -- A Network Analysis Program," IEEE Trans. on Circuit Theory, Vol. CT-20, No. 6, November 1973, pp. 628-634.

[WHI83] J. White and A. Sangiovanni-Vincentelli, "RELAX2: A New Waveform Relaxation Approach for the Analysis of LSI MOS Circuits", International Symposium on Circuits and Systems, Newport Beach, May 1983.

[WHI84] J. White and A. Sangiovanni-Vincentelli, "RELAX2.1 - A Waveform Relaxation Based Circuit Simulation Program" Custom Integrated Circuits Conference, Rochester, New York, June 1984.

[WHI85A] J. White, A.L. Sangiovanni-Vincentelli, "Partitioning Algorithms and Parallel Implementation of Waveform Relaxation Algorithms for Circuit Simulation", International Symposium on Circuits and Systems, Kyoto, Japan, June 1985.

[WHI85B] J. White, R. Saleh, A. Sangiovanni-Vincentelli, A. R. Newton "Accelerating Relaxation Algorithms using Waveform-Newton, Step Refinement and Parallel Techniques," International Conference on Computer-Aided Design, Santa Clara, CA, Nov. 1985.

[WHI85C] J. White, "The Multirate Integration Properties of Waveform Relaxation, with Application to Circuit Simulation and Parallel Computation", Ph.D. dissertation, University of California, Berkeley, ERL Memo. No. UCB/ERL 85/90, Nov. 1985.

[WHI86] J. White and A. Sangiovanni-Vincentelli, **Relaxation Techniques for the Simulation of MOS VLSI Circuits**, Kluwer Academic Publishers, Boston, MA., 1986.

[YAN80] P. Yang, I.N. Hajj and T.N. Trick, "SLATE: A Circuit Simulation Program with Latency Exploitation and Node Tearing", International Conference on Circuits and Computers, October 1980.

[YAN83] P. Yang, B.D. Epler, P.K. Chatterjee, "An Investigation of the Charge Conservation Problem for MOSFET Circuit Simulation", IEEE Journal of Solid-State Circuits, Vol. SC-18, No.1, pp.128-138, Feb. 1983.

INDEX

ABOUT THE AUTHORS

Resve A. Saleh obtained his B. Eng. Degree (Electrical) from Carleton University, Ottawa, Canada, in 1979, and his M.S. and Ph.D. degrees from U.C. Berkeley in 1983 and 1986, respectively. He has worked in industry for Mitel Corporation (Kanata, Ontario, Canada), Tektronix (Beaverton, OR), Toshiba Corporation (Kawasaki, JAPAN) and Shiva Multisystems (Menlo Park, CA). He joined the University of Illinois in 1986 where he is currently an Assistant Professor directing research in mixed-mode simulation and parallel processing. His research interests also include analog CAD and synthesis. He has served on the technical committees of the Custom Integrated Circuits Conference and the Design Automation Conference since 1987, and was a member of the organizing committee of the MidWest Symposium on Circuits and Systems in 1989.

A. Richard Newton received the B. Eng. (elect.) and M. Eng. Sci. degrees from the University of Melbourne, Melbourne, Australia, in 1973 and 1975, respectively, and the Ph.D. degree from the University of California, Berkeley, in 1978. He is currently a Professor and Vice Chairman of the Department of Electrical Engineering and Computer Sciences, University of California, Berkeley. He was the Technical Program Chairman of the 1988 ACM/IEEE Design Automation Conferences, and a consultant to a number of companies for computer-aided design of integrated circuits. His research interests include all aspects of the computer-aided design of integrated circuits, with emphasis on simulation, automated layout techniques, and design methods for VLSI integrated circuits. Dr. Newton was selected in 1987 as the national recipient of the C. Holmes McDonald Outstanding Young Professor Award of Eta Kappa Nu.